D1621492

Modular Remediation Testing System

AATDF Monographs

This monograph is one of a ten-volume series that records the results of the AATDF Program:

- Surfactants and Cosolvents for NAPL Remediation: A Technology Practices Manual
- Sequenced Reactive Barriers for Groundwater Remediation
- Modular Remediation Testing System
- Phytoremediation of Hydrocarbon-Contaminated Soil
- Steam Remediation of Contaminated Soils
- Soil Vapor Extraction: Radio Frequency Heating
- Subsurface Contamination Monitoring Using Laser Fluorescence
- Remediation of Firing-Range Impact Berms
- Reuse of Surfactants and Cosolvents for NAPL Remediation
- Surfactants, Foams, and Microemulsions for NAPL Removal

Advanced Applied Technology Demonstration Facility
(AATDF)
Energy and Environmental Systems Institute MS-316
Rice University
6100 Main Street
Houston, TX 77005-1892

U.S. Army Engineer
Waterways Experiment Station
3909 Halls Ferry Road
Vicksburg, MS 39180-6199

RICE

Rice University
6100 Main Street
Houston, TX 77005-1892

Modular
Remediation
Testing System

Edited by

Katherine Balshaw-Biddle
Rice University, Houston, TX

Carroll L. Oubre
Rice University, Houston, TX

C. Herb Ward
Rice University, Houston, TX

Authors
Thomas Reeves
Jonathan Miller
Paul C. Johnson

LEWIS PUBLISHERS
Boca Raton London New York Washington, D.C.

Library of Congress Cataloging-in-Publication Data

Catalog record is available from the Library of Congress.

Although the information described herein has been funded wholly or in part by the United States Department of Defense (DOD) under Grant No. DACA39-93-1-0001 to Rice University for the Advanced Applied Technology Demonstration Facility for Environmental Technology Program (AATDF), it may not necessarily reflect the views of the DOD or Rice University, and no official endorsement should be inferred.

Foreword

This monograph demonstrates the experience gained in design, construction, transportation, operation, and disassembly of the Advanced Applied Technology Development Facility (AATDF) modular test facility. Data on design and operation of the system are included to assist potential users or developers of similar test facilities.

An Experimental Controlled Release System (ECRS) concept was part of the original AATDF proposal to the Department of Defense, the sponsoring agency. In its initial form, ECRS was intended to be an in-ground test release site. Evolution of ECRS design into a portable system of modular components resulted from conversations and meetings with a number of advisors. The final well-sealed, above-ground modular system design was successful in avoiding leaks to the environment, facilitating mass balances, controlling test conditions, and allowing multiple uses with minimal permitting.

The people and organizations who served in an advisory capacity to AATDF during development of the ECRS concept are listed below.

- Dr. Paul C. Johnson, Arizona State University
- Drs. John W. Keeley and John Cullinane, USAE Waterways Experiment Station
- Mr. Thomas Reeves, P.E., Groundwater Services Inc. (now with Nofsinger Inc.)
- Mr. David Smyth, University of Waterloo
- Mr. Jonathan T. Miller, Equilon (formerly Shell Development Co.)
- Dr. Lynton Dicks, Shell Development Co. (retired)
- Lt. Col. Mark Smith, AF Armstrong Laboratory, and Dr. Mark Noll, Groundwater Remediation Field Laboratory, Dover AFB
- Dr. Richard Johnson, Oregon Graduate Institute
- Dr. Michael Barcelona, University of Michigan
- Messrs. Dick Brenner, John Martin, and Dan Sullivan, EPA RREL, and Al Galli, EPA Headquarters
- Dr. Corale L. Brierley, Vista Tech, and Mr. Steven J. Banks, BCM Technologies Inc.
- Groundwater Services, Inc.
- Equilon (formerly Shell Development, Westhollow Technology Center)
- AATDF Staff

The ECRS research was funded by a grant to Rice University from the Department of Defense, and managed by the U.S. Army Engineer, Waterways Experiment Station. The Experimental Controlled Release System project was led by AATDF Project Manager, Dr. Katherine Balshaw-Biddle, who also served as the senior editor for the monograph. Design development was facilitiated by the following individuals:

- Dr. Paul C. Johnson, Arizona State University
- Mr. Thomas Reeves, P.E., Groundwater Services Inc. (now with Nofsinger Inc.)
- Mr. Jonathan T. Miller, Equilon (formerly Shell Development Co.)
- Dr. Lynton Dicks, Shell Development Co. (retired)
- Dr. Katherine Balshaw-Biddle, Rice University

The editors wish to thank Ms. Mary Cormier for her assistance in preparation of the text and tables for this monograph, and Mr. Richard Conway, Union Carbide, and Dr. Karen Duston, Rice University, for their expertise in technical editing.

Preface

Following a national competition, the Department of Defense (DOD) awarded a $19.3 million grant to a university consortium of environmental research centers led by Rice University and directed by Dr. C. Herb Ward, Foyt Family Chair of Engineering. The DOD Advanced Applied Technology Demonstration Facility (AATDF) Program for Environmental Remediation Technologies was established on May 1, 1993 to enhance the development of innovative remediation technologies for DOD by facilitating the process from academic research to full scale utilization. The AATDF's focus is to select, test, and document performance of innovative environmental technologies for the remediation of DOD sites.

Participating universities include Stanford University, The University of Texas at Austin, Rice University, Lamar University, University of Waterloo, and Louisiana State University. The directors of the environmental research centers at these universities serve as the Technology Advisory Board (TAB). The U.S. Army Engineer Waterways Experiment Station manages the AATDF Grant for DOD. Dr. John Keeley is the Technical Grant Officer. The DOD/AATDF is supported by five leading consulting engineering firms: Remediation Technologies, Inc., Battelle Memorial Institute, GeoTrans, Inc., Arcadis Geraghty and Miller, Inc., and Groundwater Services, Inc., along with advisory groups from the DOD, industry, and commercialization interests.

Starting with 170 preproposals that were submitted in response to a broadly disseminated announcement, 12 projects were chosen by a peer-review process for field demonstrations. The technologies chosen were targeted at DOD's most serious problems of soil and groundwater contamination. The primary objective was to provide more cost-effective solutions, preferably using *in situ* treatment. Eight projects were led by university researchers, two projects were managed by government agencies, and two others were conducted by engineering companies. Engineering partners were paired with the academic teams to provide field demonstration experience. Technology experts helped guide each project.

DOD sites were evaluated for their potential to support quantitative technology demonstrations. More than 75 sites were evaluated in order to match test sites to technologies. Following the development of detailed work plans, carefully monitored field tests were conducted and the performance and economics of each technology were evaluated.

One AATDF project designed and developed two portable Experimental Controlled Release Systems (ECRS) for testing and field simulations of emerging remediation concepts and technologies. The ECRS is modular and portable, and it allows researchers at their sites, to safely simulate contaminant spills and study remediation techniques without contaminant loss to the environment. The completely contained system allows for accurate material and energy balances.

The results of the DOD/AATDF Program provide DOD and others with detailed performance and cost data for a number of emerging, field-tested technologies. The program also provides information on the niches and limitations of the technologies to allow for more informed selection of remedial solutions for environmental cleanup.

The AATDF Program can be contacted at: Energy and Environmental Systems Institute, MS-316, Rice University, 6100 Main, Houston, TX, 77005, phone 713-527-4700; fax 713-285-5948; e-mail <eesi@rice.edu>.

The DOD/AATDF Program staff includes:

Director:
  Dr. C. Herb Ward

Program Manager:
  Dr. Carroll L. Oubre

Assistant Program Manager:
  Dr. Kathy Balshaw-Biddle

Assistant Program Manager:
  Dr. Donald F. Lowe

Financial/Data Manager:
  Mr. Robert M. Dawson

Publications Coordinator/Graphic Designer:
  Ms. Mary Cormier

Assistant Program Manager:
    Dr. Stephanie Fiorenza

Meeting Coordinator:
    Ms. Susie Spicer

This volume, *Modular Remediation Testing System,* is one of a ten-volume series that records the results of the DOD/AATDF environmental technology demonstrations. Many have contributed to the success of the AATDF program and to the knowledge gained. We trust that our efforts to fully disclose and record our findings will help further innovative technology development for environmental cleanup.

Katherine Balshaw-Biddle

Carroll L. Oubre

C. Herb Ward

Authors and Editors

Thomas Reeves, P.E., is currently a lead mechanical engineer with Nofsinger, Inc. He brings more than seven years of diverse engineering experience to projects including plant layout, equipment and component design, specification and evaluation, and supervision of design efforts. His design experience includes P&ID development, pressure vessel and piping, application of finite element and computational methods for component design and fluid flow, and the use of statistical modeling to predict system performance. He is also involved with design for electrically classified locations, process hazard analysis, and plant unit development and expansion.

Mr. Reeves has a B.S. degree in mechanical engineering (BSME) and a master's in mechanical engineering (M.M.E.) degree from Rice University. He was employed as a mechanical engineer with Groundwater Services Inc. during the ECRS development and served as the lead engineer for design, fabrication, and testing of the two ECRS units.

Jonathan T. Miller received his M.S. degree in civil engineering from Princeton University in 1994 and joined Shell Development Company (now Equilon Enterprises LLC) the same year to work at Westhollow Technology Center. He has been involved in modeling and design of remediation strategies for sites impacted with hydrocarbons and MTBE using a wide variety of technologies and geologic settings. His research interests include practical application of numerical groundwater flow and contaminant transport models, efficient model calibration, and application of enhanced oxygen delivery technologies such as air sparging and solid peroxigens. He is also involved in application of geographic information systems (GIS) and database tools to organize historical environmental records at refining locations.

Paul C. Johnson is an Associate Professor in the Department of Civil and Environmental Engineering at Arizona State University, Tempe, AZ. His research and teaching focus on chemical fate, mass transfer, and migration in the subsurface, with practical application to *in situ* remediation, natural attenuation processes, and risk assessment. Current and past research projects have involved field-scale studies, pilot-scale tests, large- and intermediate-scale physical models, and mathematical modeling. He is the author of numerous articles and U.S. patents, and has written guidance and screening-level models currently used by many state and federal regulatory agencies. Prior to joining the faculty at Arizona State University, Dr. Johnson was a Senior Research Engineer at Shell Oil's Westhollow Technology Center. He received his doctoral degree in chemical engineering from Princeton University.

Katherine Balshaw-Biddle is an Assistant Program Manager with AATDF at Rice University. In her capacity as project manager, Dr. Balshaw-Biddle provided managerial guidance and technical expertise for the organization, implementation, and field demonstration of several projects, including creation of the ECRS modular testing unit for remediation technologies. She was also an active participant in preparation of reports for each project. Dr. Balshaw-Biddle has a Ph.D. in geology from Rice University, and her M.S. and B.S. in geology from Michigan State University.

Prior to joining the AATDF, Dr. Balshaw-Biddle worked as a senior geologist for Exxon Production Research Co. and Law Engineering and as a research scientist for Rice University, Department of Environmental Science and Engineering. She has several publications related to environmental technologies, remediation, and sedimentology.

Carroll L. Oubre is the Manager of the AATDF program. As Program Manager he is responsible for the day-to-day management of the $19.3 million AATDF program. This includes guidance of the AATDF staff, overview of the 12 demonstration projects, and ensuring project milestones are met within budget and that complete reporting of the results is timely.

Dr. Oubre has a B.S. in chemical engineering from the University of Southwestern Louisiana, an M.S. in chemical engineering from Ohio State University, and a Ph.D. in chemical engineering from Rice University. He worked for Shell Oil Company for 28 years; his last position was Manager of Environmental Research and Development for Royal Dutch Shell in England. Prior to that, he was Director of Environmental Research and Development at Shell Development Company in Houston, Texas.

C. H. (Herb) Ward is the Foyt Family Chair of Engineering in the George R. Brown School of Engineering at Rice University. He is also Professor of Environmental Science and Engineering and of Ecology and Evolutionary Biology.

Dr. Ward has undergraduate (B.S.) and graduate (M.S. and Ph.D.) degrees from New Mexico State University and Cornell University, respectively. He also earned the M.P.H. in environmental health from the University of Texas.

Following 22 years as Chair of the Department of Environmental Science and Engineering at Rice University, Dr. Ward is now Director of the Energy and Environmental Systems Institute (EESI), a university-wide program designed to mobilize industry, government, and academia to focus on problems related to energy production and environmental protection. He is also Director of the AATDF program, a distinguished consortium of university-based environmental research centers supported by consulting environmental engineering firms to guide selection, development, demonstration, and commercialization of advanced applied environmental restoration technologies for the DOD. For the past 18 years he has directed the activities of the National Center for Ground Water Research (NCGWR), a consortium of universities charged with conducting long-range exploratory research to help anticipate and solve the nation's emerging groundwater problems. He is also Co-Director of the EPA-sponsored Hazardous Substances Research Center/South & Southwest (HSRC/S&SW), the research focus of which is on contaminated sediments and dredged materials.

Dr. Ward has served as president of both the American Institute of Biological Sciences and the Society for Industrial Microbiology. He is the founding and current Editor-in-Chief of the international journal *Environmental Toxicology and Chemistry*.

AATDF Advisors

University Environmental Research Centers

National Center for Ground Water Research
Dr. C. H. Ward
Rice University
Houston, TX

**Hazardous Substances Research Center —
South and Southwest**
Dr. Danny Reible and Dr. Louis Thibodeaux
Louisiana State University
Baton Rouge, LA

Waterloo Centre for Groundwater Research
Dr. John Cherry and Mr. David Smyth
University of Waterloo
Ontario, Canada

**Western Region Hazardous Substances
Research Center**
Dr. Perry McCarty
Stanford University
Stanford, CA

**Gulf Coast Hazardous Substances
Research Center**
Dr. Jack Hopper and Dr. Alan Ford
Lamar University
Beaumont, TX

Environmental Solutions Program
Dr. Raymond C. Loehr
University of Texas
Austin, TX

DOD/Advisory Committee

Co-Chair: Dr. John Keeley
Assistant Director, Environmental Laboratory
U.S. Army Corps of Engineers
Waterways Experiment Station
Vicksburg, MS

Co-Chair: Mr. James I. Arnold
Acting Division Chief, Technical Support
U.S. Army Environmental Center
Aberdeen, MD

Dr. John M. Cullinane
Program Manager, Installation Restoration
U.S. Army Corps of Engineers
Waterway Experiment Station
Vicksburg, MS

Mr. Scott Markert and Dr. Shun Ling
Naval Facilities Engineering Center
Alexandria, VA

**Dr. Jimmy Cornette, Dr. Mike Katona,
and Major Mark Smith**
Environics Directorate
Armstrong Laboratory
Tyndall AFB, FL

**Commercialization and Technology Transfer
Advisory Committee**

Chair: Mr. Benjamin Bailar
Dean, Jones Graduate School of Administration
Rice University
Houston, TX

Associate Chair: Dr. James H. Johnson, Jr.
Dean of Engineering
Howard University
Washington, D.C.

Dr. Corale L. Brierley
Consultant
VistaTech Partnership, Ltd.
Salt Lake City, UT

Dr. Walter Kovalick
Director, Technology Innovation Office
Office of Solid Wastes & Emergency Response
U.S. EPA
Washington, D.C.

Mr. Dick Scalf
U.S. EPA Robert S. Kerr
Environmental Research Laboratory (retired)
Ada, OK

Mr. Terry A. Young
Executive Director
Technology Licensing Office
Texas A&M University
College Station, TX

Mr. Stephen J. Banks
President
BCM Technologies, Inc.
Houston, TX

Consulting Engineering Partners

Remediation Technologies, Inc.
Dr. Robert W. Dunlap, Chair
President and CEO
Concord, MA

Parsons Engineering
Dr. Robert E. Hinchee (Originally
  with Battelle Memorial Institute)
Research Leader
South Jordan, UT

GeoTrans, Inc.
Dr. James W. Mercer
President and Principal Scientist
Sterling, VA

Arcadis Geraghty & Miller, Inc.
Mr. Nicholas Valkenburg and
Mr. David Miller, Vice Presidents
Plainview, NY

Groundwater Services, Inc.
Dr. Charles J. Newell
Vice President
Houston, TX

Industrial Advisory Committee

Mr. Richard A. Conway, Chair
Senior Corporate Fellow
Union Carbide
S. Charleston, WV

Dr. Ishwar Murarka, Associate Chair
Electric Power Research Institute (EPRI)
Currently with Ish, Inc.
Cupertino, CA

Dr. Philip H. Brodsky
Director, Research & Environmental
  Technology
Monsanto Company
St. Louis, MO

Dr. David E. Ellis
Bioremediation Technology Manager
DuPont Chemicals
Wilmington, DE

Dr. Paul C. Johnson
Arizona State University
Department of Civil Engineering
Tempe, AZ

Dr. Bruce Krewinghaus
Shell Development Company
Houston, TX

Dr. Frederick G. Pohland
Department of Civil and Environmental
  Engineering
University of Pittsburgh
Pittsburgh, PA

Dr. Edward F. Neuhauser, Consultant
Niagara Mohawk Power Corporation
Syracuse, NY

Dr. Arthur Otermat, Consultant
Shell Development Company
Houston, TX

Mr. Michael S. Parr, Consultant
DuPont Chemicals
Wilmington, DE

Mr. Walt Simons, Consultant
Atlantic Richfield Company
Los Angeles, CA

Executive Summary

The DOD-funded Advanced Applied Technology Demonstration Facility (AATDF) at Rice University developed and implemented a modular testing system for remediation technologies that facilitates controlled release of contaminants and real-time monitoring of the technology efficiency. A database on 13 test release facilities was developed, and the AATDF's Experimental Controlled Release System (ECRS) evolved from the experiences of these facilities. Detailed discussion of the ECRS design, with engineering drawings, information for shipping and setup, procedures for operations and maintenance, monitoring approaches, testing results, and safety and health procedures is presented.

Two ECRS portable units were constructed. Each was designed to facilitate cost-effective testing of technologies or processes in a three-dimensional, pilot-scale setting, with closely controlled vadose or ground-water conditions. The soil tank can be packed and instrumented to simulate a variety of subsurface conditions. The equipment and instrumentation were selected to accommodate a range of contaminants.

Each unit consists of four modules, including an instrumentation building, process equipment skid, soil tank and reservoirs. The insulated instrumentation building is climate controlled and provides space for analytical and computer equipment. The building has an exterior bottled-gas rack, stainless steel piping for gas chromatography (GC) gases, electrical outlets, bench-tops and storage space. The process equipment skid is equipped with a compressor, a blower, ground-water pumps, water and air filters, piping, an electrical panel, a gauges panel, and a data recording/display unit for real-time acquisition and monitoring of system parameters and system controls. The soil tank is a 27-yd³ rectangular steel sludge container, epoxy coated and fitted with flanges, tubing nozzles, valves, tee-strainers, and sight gauges. It has a flexible top that is sealed and bolted to the tank lip, as well as a pressure relief valve and two high-liquid level switches for safety. The system also includes water reservoirs and a chemical mixing tank in the fourth module. The equipment modules for each unit pack onto one flatbed trailer for shipping.

The two modular remediation testing systems were conceived, designed, constructed, shipped, installed, operated, and monitored to allow cost-effective, quantitative evaluation of a range of technologies in a number of site-specific scenarios. Unit 1 was set up at Shell Westhollow Technology Center, Houston, TX, in the spring of 1996. It underwent a safety inspection and an initial 3-month systems test of the equipment followed by a 12-month air sparging research project. Mr. Jonathan T. Miller, Equilon (Shell Development Co.), was the principal investigator on the project. His soil pack and sensor/sampler array were designed to test the effect of minor soil heterogeneities on air sparging. He also varied air sparging operating parameters and tested the effectiveness of oxygen releasing materials. Unit 1 was disassembled in August, 1997. It underwent minor equipment retrofitting and was shipped to the USAE Waterways Experiment Station (WES), in Vicksburg, MS, in January, 1998 to be used for long-term testing of selected remediation technologies.

Based upon the performance of Unit 1 during the air sparging research at Shell's Westhollow Technology Center and later quantitative system testing by Groundwater Services, Inc. (GSI), it was determined that the system performed to design specifications. After completing the post-performance review, the AATDF solicited comments and suggestions for changes in design or performance from the Shell researchers and advisors, the GSI fabrication team headed by Thomas Reeves, and other ECRS advisors, including Dr. Paul Johnson, Arizona State University. The main design enhancements included the following:

- addition of a sight glass column to view the water level in the tank
- substitution of a centrifugal pump for the rotary gear pumps to facilitate moving the filters downstream of the pump
- substitution of a screw-type air compressor for the centrifugal air pump

- replacement of the SitePro-SpargePro Controller System with a data controller
- location of all gauges and controls on one panel on the process equipment skid
- addition of humidity control to the instrumentation building to reduce condensation
- substituting a portable GC for a heated sample line and thermal vapor analyzer instrumentation that was removed
- replacement of the HDPE tank top with a two-ply urethane fabric top
- installation of bolts through the tank lip to secure the fabric top
- addition of an air accumulator, supply-air tank to the process equipment skid
- addition of a portable 50-gal chemical mixing tank

Unit 2 fabrication was completed in January 1997. It was shipped to Arizona State University, Tempe, AZ in the spring of 1997, as part of an air sparging project being conducted by Dr. Paul Johnson, a member of a large research team addressing various aspects of physical model studies of *in situ* air sparging performance. Funding for the larger project was provided by the USAF, SERDP, API and AATDF/Rice University. Dr. Johnson used ECRS Unit 2 to test air sparging treatment methods for source zones and dissolved plumes and to perfect diagnostic tools for air sparging performance. Unit 2 was returned to Rice University, in August 1998, for the use of university researchers testing remediation processes or technologies. It also underwent retrofitting to simplify and improve system performance.

As detailed in Chapter 7 of this monograph, the ECRS is a unique, modular testing environment. The major attributes of this system that distinguish it from past test facilities fall into the following categories:

- Portable — shippable to researcher's location or remediation site
- Tightly sealable — facilitates mass balance
- Large pilot-scale facility — step below full-scale demonstration
- Flexible testing conditions — vadose zone or aquifer, chemical release or contaminated soil, air sparging or SVE
- Easy to construct — design drawings developed, standard equipment
- Easy to operate and maintain — easy access, standard components
- Affordable — easily shipped and set up, reduced regulatory requirements, minimal maintenance
- Faster tests — reduced or eliminated permitting, operate within one week of arrival, programmable for 24-hr operation

This monograph presents the information needed by others to design, construct, and operate similar units or to utilize the existing units when available.

Acronyms and Abbreviations

AATDF	Advanced Applied Technology Development Facility
AFB	Air Force Base
API	American Petroleum Institute
AST	aboveground storage tank
ASU	Arizona State University
BTEX	benzene, toluene, ethylbenzene and xylenes
cfm	cubic feet per minute
CFR	Code of Federal Regulations
CRADA	Cooperative Research and Development Agreement
db	decibels
DI	deionized water
DO	dissolved oxygen
DOD	U.S. Department of Defense
DOE	U.S. Department of Energy
DOT	U.S. Department of Transportation
DNAPLs	dense nonaqueous phase liquids
DPDS	n-hexadecyl diphenyloxide disulfonate
PDG	differential pressure gauge
ECRS	Experimental Controlled Release System
EPA	U.S. Environmental Protection Agency
EMSL	EPA Environmental Monitoring Systems Laboratory, Las Vegas, NV
ESF	Environmental Simulation Facility, University of Wyoming, Laramie, WY
E-TEC	EPA Engineering and Technology Evaluation Center, Edison, NJ
GAM	granular activated material
gpm	gallons per minute
GC	gas chromatograph
GC/MS	gas chromatograph/mass spectrometer
GPR	ground penetrating radar
GRFL	A.F. Armstrong Lab's Groundwater Remediation Field Laboratory, Dover AFB
GRI	Gas Research Institute
GSI	Groundwater Services, Inc.
HDPE	high density polyethylene
hp	horsepower
HSRC	Hazardous Substance Research Center
HVAC	heating, ventilation, and air conditioning
IAS	*in situ* air sparging
LAS	reference surfactant
MBAS	methylene blue active substance
MSDS	Material Safety Data Sheet
MTBE	methyl tertiary butyl ether
NAPLs	nonaqueous phase liquids
NAS	Naval Air Station
NEMA	National Electric Manufacturer's Association
NETTS	National Environmental Technology Test Site
NFEC	Naval Facilities Engineering Command
NFESC	Naval Facilities Environmental Service Center
NCIBRD	National Center for Integrated Bioremediation R&D
NERL	U.S. EPA National Exposure Research Laboratory

NJIT	New Jersey Institute of Technology
NRMRL	National Risk Management Research Laboratory
NMSU	New Mexico State University
OD	outer diameter
OGI/LEAP	Oregon Graduate Institute/Large Experimental Aquifer Program, Beaverton, OR
ORC	oxygen-releasing compound
ORMs	oxygen-releasing materials
OSHA	Occupational Safety and Health Administration
PAHs	polynuclear aromatic hydrocarbons
PCE	perchloroethylene
pcf	pounds per cubic foot
PID	proportional, integral, and digital
POL	petroleum oils and lubricants
PPE	personnel protective equipment
ppb	parts per billion
ppm	parts per million
psi	pounds per square inch
PVC	polyvinyl chloride
RH	relative humidity
RREL	U.S. EPA Risk Reduction Environmental Laboratory, Cincinnati, OH (now NRMRL)
RSKERL	EPA R.S. Kerr Environmental Research Laboratory, Ada OK (now part of NRMRL, Subsurface Protection and Remediation Division)
scfm	standard cubic feet per minute
SERDP	Strategic Environmental Research & Development Program
SITE	Superfund Innovative Technology Evaluation
SOP	standard operating procedure
SVE	soil vapor extraction
TNRCC	Texas Natural Resources Conservation Commission
USAE	U.S. Army Engineers
USAF	U.S. Air Force
UST	underground storage tank
VEGAS	Versuchseinrichtung zur Grundwasser und Altlastensanierung, Stuttgart, Germany
WCS	Waste Control Specialists
WES	U.S. Army Engineer Waterways Experiment Station, Vicksburg, MS
WTC	Shell Oil, Westhollow Technology Center, Houston, TX

Contents

List of Photos in Text

List of Figures in Text

List of Tables in Text

List of Figures in the Appendices

CHAPTER 1

Introduction and Technology Overview

1.1 PROJECT BACKGROUND

The DOD-funded Advanced Applied Technology Demonstration Facility (AATDF) at Rice University developed and implemented a remediation technology testing system designed to facilitate controlled release of contaminants, mass balance calculations, and real-time monitoring of cleanup efficiency. The Experimental Controlled Release System (ECRS), as developed by AATDF, now consists of portable, modular testing equipment and instrumentation that can be shipped to a researcher's location or a remediation site for pilot-scale treatability or technology testing.

During preparation of the AATDF funding proposal by Dr. C. Herb Ward and others, Dr. John Cherry (University of Waterloo) suggested that the proposal include construction of a test release site in the U.S. It would serve as a field test site for remediation technology development, similar in function to the University of Waterloo site at Canadian Forces Base (CFB) Borden, Ontario, Canada.

Following the award of the AATDF Program to Rice University, the USAF Armstrong Laboratory, with support from the Strategic Environmental Research and Development Program (SERDP), decided to build an experimental controlled release site for testing remediation technologies based upon a design similar to that at CFB Borden. Rice University signed a memorandum of understanding to work cooperatively with the Air Force on design of the site. The site selected by the Armstrong Laboratory was located at Dover AFB in Delaware and named the Groundwater Remediation Field Laboratory, one of the SERDP-sponsored National Environmental Technology Test Sites (NETTS). AATDF and its advisors decided that the AATDF Program could still add value to the national environmental technology development and evaluation effort by building an innovative test system concept.

The AATDF staff then gathered information about existing test release facilities, potential locations and partners, designs, costs, and regulatory requirements. Existing or planned facilities that provided information to the AATDF about equipment, operations, and costs, or proposed designs, included the Waterloo Centre for Groundwater Research, Oregon Graduate Institute's Large Experimental Aquifer Program, the U.S. Air Force Armstrong Laboratory's Groundwater Remediation Field Laboratory (GRFL), the U.S. Army Engineer Waterways Experiment Station (WES) Laboratory, EPA DOE Laboratories, Center for Environmental Simulation Studies at the University of Wyoming, the VEGAS facility (Germany), the Colorado School of Mines' Rocky Flats Local Initiative, National Center for Integrated Bioremediation Research and Development (NCIBRD), and Waste Control Specialists (WCS). Some of the information provided by those facilities is briefly discussed in Section 1.2.

Partnering with a developing or existing facility was considered as a means for leveraging the AATDF funding. Exploratory meetings were held with interested parties and AATDF advisors from the Armstrong Laboratory, WES, EPA, WCS, and the ECRS Advisory Committee. A draft business plan was also developed with the assistance of two AATDF commercialization advisors. Due consideration was then given to data gathered on costs, potential locations, regulatory review and requirements, research uses, and projected schedules for site construction and technology testing. It was decided that construction, regulatory permitting, and maintenance of a test release site would be costly in terms of both time and funding. The preferred option was to build a portable, pilot-scale testing unit that could be shipped directly to a researcher. This option was also expected to appeal to researchers because it eliminated travel costs to distant field sites, allowed use of their full research staff, and provided for a longer period of on-site monitoring for tests.

After finalizing this change in the ECRS concept from a test site to a portable testing system, the AATDF staff and ECRS advisors focused on designing durable, modular units of equipment that could be shipped and assembled with relative ease and could also be instrumented and tightly sealed to facilitate mass balance measurements. The design is discussed in Section 2.0.

1.2 TEST RELEASE FACILITIES DATABASE

Information was gathered from more than 13 operating or proposed test release facilities during the period 1994–1995. Many of the facilities are listed below (in alphabetical order) and described briefly in Table 1.1.

- Air Force Armstrong Laboratory, Groundwater Remediation Field Laboratory, Dover AFB, DE
- U.S. Army Engineer Waterways Experiment Station, Vicksburg, MS
- Environmental Simulation Facility (ESF), University of Wyoming (Dept. of Range Management), Laramie, WY
- U.S. EPA R.S. Kerr Environmental Research Laboratory (RSKERL), Ada, OK
- U.S. EPA RREL Testing and Evaluation Laboratory, Cincinnati, OH
- U.S. EPA RREL Releases Control Branch, Edison, NJ
- U.S. EPA Engineering and Technology Evaluation Center (E-TEC), Edison, NJ
- U.S. EPA Environmental Monitoring Systems Laboratory, Las Vegas, NV
- U.S. DOE Lawrence Livermore Laboratory, Livermore, CA
- National Center for Integrated Bioremediation Research and Development (NCIBRD), Wurtsmith AFB, Oscoda, MI
- Oregon Graduate Institute/Large Experimental Aquifer Program (OGI/LEAP), Beaverton, OR
- Rocky Flats Local Initiative, Institute for Resource and Environmental Geosciences, Colorado School of Mines, Golden, CO
- Versuchseinrichtung zur Grundwasser und Altlastensanierung (VEGAS, Experimental facility for groundwater and existing waste sites), Stuttgart, Germany
- Waste Control Specialists, Pasadena, TX
- Waterloo Centre for Groundwater Research, Waterloo, Ontario, Canada

The types of information requested from the test release facilities are listed below. Not all facilities could provide this level of information at the time, but most were very helpful in providing a comprehensive look at their research mission, equipment design, and general costs. Some of the facilities were dedicated for test release projects, others were laboratories with test release capacity, and still others were in the planning stage and may or may not have been developed later. The requested information included the following categories:

- management infrastructure to administer site
- number and sizes of test cells or tanks
- typical construction of cells or tanks
- typical schedule for use

- types of technologies tested
- types of heavy equipment used at site
- other equipment and facilities (such as an on-site analytical laboratory)
- design of groundwater injection system in tanks or groundwater flow control in cells
- expected lifetime of cells or tanks
- management of waste soils and wastewater after test
- methods for testing integrity of seals (in walls, around cassettes, in tops)
- types of permits required for facility and prior to release
- funding vs. costs
  fees charged to outside users
  annual maintenance costs
  estimated costs for monitoring, sampling, analysis, waste treatment, waste disposal
- information on sheet pile cells with clay bottom
  installation costs
  methods to test permeability of clay unit
  methods to map clay surface to determine depth for sheet piling and regional slope
  methods to determine if clay seals base of sheet piles
  other materials for *in situ* cell walls besides sheet piles
  testing sealing integrity of grout in the sheet piling joint

A summary of data received from existing and proposed facilities with test release capabilities is presented below. The completed database was used both as a market survey to help define a specialized niche for the ECRS, and also as an engineering resource for the ECRS design. Although some of the information may no longer be current, it is presented here to illustrate the database that was used during the ECRS development and to provide design ideas for various situations. Table 1.1 summarizes key features.

GRFL, Dover AFB, DE, is an in-ground test release facility that was constructed in 1995–1996. At the request of state regulators, Armstrong Laboratory prepared an Environmental Assessment for the site and received a state permit for the facility. Testing is performed in double walled sheet-pile cells ranging 36 to 1800 ft^2 in size. The sheet pilings are driven 30 to 40 ft into the subsurface and penetrate 3 to 5 ft into an underlying clay aquitard. The pilings are connected laterally by joints that are tremie-sealed (vibrated into place) with bentonite grout in a manner developed by the University of Waterloo. The annular space between the inner and outer sheet piling cells is filled with water to provide an inward hydraulic gradient. Monitoring wells are installed in the annular space and outside the perimeter of each test cell. The top of the soil in the cell is sealed with a polyethylene vapor barrier overlain by approximately 6 in. of sand, and the entire cell with instrumentation and monitoring equipment is covered with a portable building (fabric over metal frame). The GRFL has an on-site manager and its own analytical and office facilities housed in a portable building at the site.

WES, Vicksburg, MS, routinely performs test release and remediation or treatability studies of contaminants in a variety of containerized systems, ranging in size from bench-scale to pilot-scale. Containers used for these studies include lysimeters, sludge tanks, columns, and metal trolleys. WES has a state-of-the-art analytical facility on site, in addition to full design, fabrication, and engineering support.

ESF, University of Wyoming (Department of Range Management), Laramie, WY, was planned for construction on campus and modeled after a prototype facility at the university. The plans included large, portable lysimeters, $20 \times 24 \times 10$ ft deep, constructed within environmental chambers that simulate daylight, rainfall, stream flow, and groundwater aquifer conditions. Two lysimeters with sampling ports would fit inside one concrete tank and under one environmental chamber cover. The lysimeters would be constructed of concrete, insulated on the interior with a sheet of foam and lined with a durable, chemical resistant liner. One end would be constructed of stainless steel, with evenly spaced portals for sampling and a Plexiglas window for viewing. Overhead loading and unloading of soil and equipment from chambers would be accomplished on-site or off-site.

The prototype facilities in the Department of Agriculture at UW were used for research related to agriculture or oil shale, but ESF would focus on testing remediation processes and technologies.

RSKERL, in Ada, OK (now the *National Risk Management Research Laboratory (NRMRL), Subsurface Protection and Remediation Division)*, has large-scale model aquifers built in the mid-1980s and used to evaluate transport and fate of dense nonaqueous phase liquids (DNAPLs). There are two indoor tanks, each $4 \times 15 \times 4$ ft, constructed with an angle-iron frame and double-lined with Plexiglas and glass. In 1995, one tank was empty for repair because internal pressure on the inner glass liner caused it to crack. There were plans to convert to stainless steel for future aquifer models.

The criteria used to determine the size of this test aquifer included the total volume of water sampled for laboratory analyses, the number and volume of soil samples, the groundwater volume and velocity, and the boundary effects near the tank walls and lack of steady-state flow in the saturated capillary fringe. Flow rises vertically in the capillary fringe near the injection point and drops vertically near the recovery point.

One of the model aquifers, used for surfactant research on perchloroethylene (PCE), was packed with three sediment layers: sandy surface soil with organic materials, clean quartz sand, and clay. Air-dried soil was poured in thin lifts with minimal vertical drop, to prevent separation of any fines. Successive lifts of the same grain-size soil were homogenized by inserting a knife through the lifts and stirring. The soils for this test were collected near Ada, OK, and included clay from a brick plant (a mixture of bentonite, kaolinite, and illite), quartz sand from a local quarry, and sandy surface soil with roots and other organic debris.

Water was injected through vertical stainless steel screens at one end of the model, collected in the same manner at the other end, and recirculated, with contaminant, throughout the experiment. The researcher recommended using a displacement pump instead of constant head to inject water because biological growth in the media near the injection point interfered with constant head injection. The sediment grain-size distribution in the tank was modeled as a geostatistical distribution (single element finite grid) of the three dimensional soil distribution at one site at a DOD base. The grid system was maintained by placing sediments in brick-size cells with 4-in. walls.

EPA RREL Testing and Evaluation Laboratory, Cincinnati, OH (now *NERL Microbiological and Chemical Exposure Division)* had 25,000 ft^2 of space for testing hazardous waste soil treatments at the converted (former) city wastewater treatment plant. The facility had 400 to 500 gal lysimeter tanks, which were dismantled but could be rebuilt, 55-gal drum composters, 60-liter bioslurry reactors, grinders, and screens for soil preparation, and was planning to install two IDEA climate control chambers with lysimeter walls that could hold 3 to 4 ft of soil in its base. The facility had RCRA Part A and B permits and was operated by International Technologies (IT), Inc. for the EPA.

EPA RREL Releases Control Branch, Edison, NJ (now *NRMRL Water Supply and Resources Division)* had an underground storage tank (UST) testing research facility that included a pea-gravel filled tank hold containing UST-equipped and monitoring devices for leaks.

E-TEC, Edison, NJ, had been planned at the former Raritan Arsenal, now an industrial park and community college in New Jersey, but was not built due to increasing costs related to regulatory requirements for waste treatment. The intent of the facility was to test EPA SITE Program technologies to the point of failure. It was to have been operated in conjunction with the New Jersey Institute of Technology (NJIT) and would have been housed in two large refurbished warehouses on the site. Plans had been completed for three testing bays in two buildings to be used for *ex situ* testing, an analytical laboratory, library, and office. Testing was to be performed in trailer-mounted tanks that could easily be rolled into the testing bays for treatment and then removed for soil disposal. The facility was proposed during the 1980s when the SITE program was having difficulty finding field sites to test remediation technologies.

EPA Las Vegas, Environmental Monitoring System Laboratory (now *NERL Environmental Sciences Division)* was considering building a test release facility near Las Vegas, NV, to perform geophysical research on resistive tomography, seismic tomography, bore-hole ground penetrating radar (GPR), electromagnetic surveys, and spectral induced polarization. They were at the stage of considering sites, and no design information had been developed.

U.S. DOE Lawrence Livermore Laboratory, in Livermore, CA, used tanks for high pressure research. Information on coating materials for tank interiors was provided.

NCIBRD, Wurtsmith AFB at Oscoda, MI, was developed by EPA Superfund's Great Lakes and Mid-Atlantic Hazardous Substance Research Center with support from the DOD and the EPA. The facility is one of the testing locations for the DOD's NETTS, and is associated with the University of Michigan, Department of Civil and Environmental Engineering. Contaminants at the 43 sites available for research at Wurtsmith include NAPLs, chlorinated solvents, heavy metals, and poly-nuclear aromatic hydrocarbons (PAHs). Contaminated media at the sites include surface water and bottom sediments, vadose soils, groundwater, and aquifer soils. A hydrogeochemical database has been developed for Wurtsmith based on over 600 borings and wells. An original test design at Wurtsmith was an uncontained *in situ* release monitoring system for groundwater plumes. A contained test cell was installed later.

The *Oregon Graduate Institute/Large Experimental Aquifer Program (OGI/LEAP),* Beaverton, OR, built its first outdoor tank in 1988. Research at LEAP has included study of gasoline and solvents in the vadose zone, monitoring transport and fate in groundwater, testing site character-ization technologies, monitoring equipment, and remediation technologies. The first tank was built for an EPA-funded project simulating gasoline behavior in UST backfill (pea gravel and washed sands). It was designed to simulate a tank-hold environment and consisted of a $10 \times 10 \times 3$ m unlined concrete bunker. The largest current tank is an in-ground $32 \times 70 \times 15$ ft deep, mild carbon steel tank with a central steel divider. Other tanks include three large in-ground tanks with dimensions of $10 \times 10 \times 5$ m and two medium above-ground tanks with dimensions of either 8 or 10 m $\times 2.5 \times 0.5$ m.

Tank construction consisted of mild carbon steel with a 10-in. thick concrete outer liner containing double rebar, a 6-in. annular space filled with pea gravel and water, an inner steel bunker framework of angle- and I-beams, and a steel floor underlain by pea gravel. The tank cover is constructed from polyvinyl chloride (PVC) sheets, taped and glued around monitoring points. Groundwater injection occurs through 6-ft screens in three vertical wells, which are connected at the base by a common screen. Water quality in the annular space is monitored at set intervals during the experiments. LEAP also uses assorted cubic-meter boxes and column studies in the laboratory. Analytical work is done on-site at OGI laboratories.

Transport and fate studies in the big tank take about 2 years. Heavy equipment includes a track hoe and a bob cat. Soil lifts of sand and gravel are placed in the tank dry, saturated with water, then drained to improve compaction; bentonite is poured in dry and covered with sand before being saturated with water and drained. Monitoring equipment is installed after soil is in place. Filling one side of the large tank takes a week. Small diameter (0.5-in.) stainless steel sampling bundles, glass video observation tubes, pressure transducers (water level), thermisters (temperature), data acquisition system may be installed.

In chlorinated tests, the sediments are layered by hand without a mechanical compactor. As the tank is filled with layers of sand, clay (dry bentonite or silica flour), and gravel, it is flooded and drained a few times to compact the soils. Porosity in the hand-packed tank measures around 35%. The greatest compaction achieved in the sediment pack using the flooding technique was about 1 cm over 4 m-thick pack.

Rocky Flats Local Initiative, Institute for Resource and Environmental Geosciences, Colorado School of Mines, Golden, CO, was planned as a consortium of universities and consulting companies to provide research on remediation technologies for Rocky Flats contaminants and other DOE sites. The design included construction of three above-ground tanks installed above floor level in a building. The largest tank was $8 \times 12 \times 4$ m deep; the second was $4 \times 6 \times 2$ m deep; and the third was $2 \times 3 \times 1$ m deep.

Tank construction was to consist of concrete with a double liner to allow use of geophysical characterization and monitoring tools. Stainless steel tubes were to be mounted through the tank walls and floor for sampling soil gas and liquids, and monitoring temperature, chemical concentrations, and tracers. Glass monitoring wells were planned for video and fluorescent viewing. A hinged side on the tank would allow sediment input and excavation with a backhoe assisted by an overhead crane. The tanks were to be housed in an existing warehouse building near Rocky Flats, and positioned over catch basins in the floor for leak protection. Contaminated soil was to be disposed of or incinerated at a facility in Colorado.

VEGAS - Versuchseinrichtung zur Grundwasser und Altlastensanierung (experimental facility for groundwater and existing waste sites) is located at the University of Stuttgart, Vaihingen campus, Stuttgart, Germany. Its mission includes developing techniques for *in situ* remediation of contaminated soil and groundwater, testing mobility of contaminants in the subsurface, improving methods for site characterization and determining contaminant mass. The program relies on a consortium of ten German universities, seven research institutes, nine industrial companies, one government research facility and four international research groups (Waterloo, UC Berkeley, two French institutes). Construction was completed in 1995, and the facility houses several large-scale stainless steel tanks ranging up to 850 m^3 in volume. The largest tank is $19.5 \times 9 \times 4.5$ m and can be subdivided into three equal-sized cells. A second tank is $6.5 \times 2.2 \times 2.5$ m. Other tanks include a 3-m diameter lysimeter and two testing flumes ($16 \times 1 \times 3$ m and $10 \times 0.2 \times 0.7$ m). The tanks are housed in a large open-format building ($36 \times 18.5 \times 17$ m deep) that has an interior overhead crane with a loading capacity of 5 tons. The building is being equipped with high voltage electrical supply, compressed air, nitrogen, and water outlets. A 150 m^2 area within the building houses smaller test containers.

The water treatment system has a capacity of 50 m^3/day, with two temporary storage tanks each holding 50 m^3 of water. Volatiles and exhaust gases are processed through a vacuum exhaust cleaning system. Other available laboratories on the premises include climate control room, explosion protection room, contaminated material storage and loading room, analytical laboratory, testing and electronic laboratory, offices, and changing room.

The *WCS Anderson County hazardous waste landfill* site in west Texas was planned to include a technology testing facility equipped with two in-ground tanks and a sheet piling cell. Maximum tank depth was 15 ft to accommodate use of a trackhoe for soil removal. Tank construction would be double-walled carbon steel. The excavation tank hole would extend at least 5 ft into the Triassic Red Bed clay soil and have a 1% slope to a sump to collect DNAPLs. The bottom would have a liner overlain by sand or gravel with a secondary leak detection system. Tanks would be sealed with a geomembrane and housed in portable greenhouse structures with temperature control and thermostatic fans. Plumbing for groundwater injection and withdrawal would use a horizontal, oversized, perforated manifold with gravel pack to minimize biological plugging activity at the injector. DNAPLs would be collected at the withdrawal end of the tank. Analytical facilities would be available on site.

The *Waterloo Centre for Groundwater Research*, Waterloo, Ontario, Canada, was the first sheet-pile test release center in North America. It is located at Canadian Forces Base Borden, and testing is done in patented sheet-pile cells driven into a clay aquitard at 40 ft. Waterloo researchers are currently using the site to develop funnel and gate technology with treatment cassettes. The contaminants of interest have been BTEX and solvents.

Sheet piling cells are constructed using patented Waterloo sheet pilings with interlocking grouted joints at a cost of approximately $25/ft^2. The site has its own small drill rig and hand-operated portable equipment to take soil cores and install groundwater monitoring devices. They use commercial contractors for larger drilling tasks and subcontract all heavy machinery. The site has several trailers, three winterized portable structures, three small sheds, one large winterized building (10×10 m), four residential campers, and a washroom trailer.

Routine analytical work is done at Waterloo, a 2-hr drive from the Borden site. Portable equipment used at the site includes several gas chromatographs (GCs), a gas chromatograph/mass spectrometer (GC/MS), and other analytical devices and meters. Utilities are supplied by the base and paid by Waterloo, and these costs are included in the annual site and equipment maintenance budget.

Waterloo has run up to ten experiments in sheet-pile cells during a one-year period at the site (1991 and 1992). In mid-1994, five experiments were operated in sheet-pile cells ranging in size from 1.5 × 4.5 m to 9 × 9 m over depths of 2.5 to 14 m and two funnel-and-gate structures. During 1996–1997, the Borden AATDF work was performed in three parallel sheet-pile cells with removable treatment cassettes.

Test release projects have been undertaken without containment cells at the following sites:

Columbus AFB, MS: The MADE 1 and 2 Tracer Studies involved release of a tracer and dissolved hydrocarbon fuel. There were 3000 sampling points installed to monitor the release and tracer study. A release of simulated jet fuel had been planned for a future study.

New Mexico State University, Las Cruces, NM: A trench experiment was performed to measure contaminant transport *in situ* and predict simulation of contaminant transport in Rocky Flats tests beds. NMSU is a member of the Business-University Test Bed Consortium that manages the Rocky Flats project.

Stanford University, Palo Alto, CA: Researchers at Stanford performed test release studies at NAS Moffett Field, CA, where soluble-phase aromatic chemicals and chlorinated solvents were released in the well pattern for bioremediation research. They also performed test release studies using BTEX at Seal Beach, CA.

1.3 DEVELOPMENT OF THE ECRS

Mission Statement

After gathering available data on existing and planned test release facilities, a draft ECRS mission statement was written to focus development to best satisfy DOD needs for remediation technology testing.

The Experimental Controlled Release System (ECRS) is being developed to facilitate controlled release of contaminants and chemical amendments for quantitative technology testing. The researcher will be able to monitor equipment performance, calculate mass and energy balances, and test the clean up efficiency for innovative remediation concepts and technologies. The ECRS design will promote demonstration of innovative processes and technologies under tightly controlled conditions, using best scientific and engineering principles and practice.

ECRS testing will focus on selected DOD contaminants of interest including fuels, solvents, petroleum oils and lubricants, heavy metals, and mixtures of these wastes. The media of interest include soil and groundwater. The technologies and processes of interest include *in situ* contaminant destruction or mobilization, plus enhanced site characterization/monitoring tools.

A main objective is accurate measurement of contaminant removal efficiency through calculation of mass and energy balances. To this end, the ECRS design will be a closed system and researchers will have access to portable analytical equipment, such as a gas chromatograph and a data logger. This instrumentation will be housed in a portable climate-controlled building where researchers will also be able to use or store their own instrumentation for sampling and analysis.

After completion of the mission statement and the database of test release facilities, AATDF convened the ECRS Advisory Committee to discuss this information. The committee membership included representatives from some of the existing and planned test release facilities, in addition to researchers who had experience with remediation technology testing either using these testing facilities or other laboratory and field sites. The membership of the committee is given in the Foreword.

The participants outlined their experiences regarding test release projects and the facilities listed in Section 1.2 and presented in Table 1.1. The members were also asked to provide suggestions on a number of topics related to ECRS development, which are listed below. Potential ECRS designs considered during the meeting focused on sheet-piling cells, landfill cells with a synthetic liner, and permanent tanks (in-ground and above-ground tanks).

ECRS Advisory Committee meeting agenda:

- Test Release Facilities
  History
  Existing facilities
  Proposed facilities
- Philosophy
  Types of technology projects suited to a test release facility
  Project duration, typical vs. optimal schedule for tests
  Methods to track and optimize mass and energy balances
  Site selection/facility design criteria
  Mission statement
- Design Criteria
  Cell design options
  Cell size
  Flexibility for multiple technology types
  Soil packing methods for tanks
  Cell matrix/contaminant characterization
  Monitoring equipment
- Site Construction
  Construction criteria
  Regulatory approval and permitting
  Time schedule
  Costs
- Site Operation and Maintenance
  Management personnel
  Management philosophy
  Facility staff
  Analytical equipment
  Heavy equipment
  Office space
- Technology Transfer and Commercialization Options
- 5- to 10-Year Plan
  Future users and funding sources
  Future technology testing

Selected comments and design suggestions arising from the ECRS Advisory Committee meeting are summarized in Table 1.2.

Potential Sites and Partners

The AATDF also explored options for potential partners and sites for the ECRS through its DOD Advisory Committee and other contacts. Data were collected for a variety of potential sites and a range of partnering options. Organizations initially considered for potential partnering on ECRS development included industry-funded research centers, DOD organizations, an EPA laboratory and regional Hazardous Substance Research Center (HSRC), and a DOE laboratory.

Table 1.1 Summary of Planned and Operating Test Facilities in 1994–1995 (listed alphabetically by acronym)

	Stage of Development	Location	Funding	Facility Design	Contaminants of Interest	Media of Interest	Technology Categories to be Tested	Technology Tests	Research Goals
ESF Environmental Simulation Facility, Univ. of Wyoming	Design	Laramie, WY	Possible state or private	Portable lysimeters in environmental chambers	Not specified	Soil	Not specified	Based on client needs	
ECRS Experimental Controlled Release Site	Planning/design phase	Units at Shell and ASU then at WES and Rice	DOD	27-ft³ portable test cells and operating equipment to be shipped to researcher (used outdoors under supplied shelter)	DOD concerns: fuels, solvents, munitions, heavy metals	Soil and groundwater	*In situ* destruction, mobilization, characterization	Surfactants, site characterization tools (tracers, etc.), others	Mass and energy balance, technology scale-up, design manuals
EPA-RSKERL R.S. Kerr Envi. Research Lab (EPA NRMRL)	Operating	Ada, OK	EPA SERDP	2 large-scale "model aquifers," iron frame w/ plexiglass; will build steel 3-D or 2-D model	Solvents, fuels	Soil and groundwater	*In situ* bioremediation, mobilization	Surfactants, site characterization tools (tracers, tomography) bioventing	Transport and fate, bioremediation
EPA RREL	Operated by IT for EPA	Cincinnati, OH	EPA	500-gal lysimeters, reactors, composters, planning to install two IDEA climate control chambers w/lysimeter walls	Hazardous waste (fuels, POL, solvents, metals)	Soil	Bioremediation, destruction, mobilization		
E-TEC EPA Engineering and Technology Evaluation Center	Design phase	Edison, NJ	Proposed EPA SITE Program funding	Tanks mounted on semi-truck trailers driven into warehouse bay testing space, on site analytical lab and offices	SITE Program contaminants of concern	Soil and hazardous wastes	Contaminant destruction or removal	SITE Program technologies	Test technology to the point of failure
GRFL A. F. Armstrong Lab Groundwater Remediation Field Lab	Planning/design phase (constructed 1995–1996)	Dover AFB, Dover DE	SERDP	Field demos in sheet piling cells, analytical lab on site	DOD National Testing Site for solvents research; fuels	Soil and groundwater	*In situ* destruction, mobilization, solidification	Based on SERDP needs	Development of remediation technologies for AF cleanup/ compliance
OGI/LEAP OGI/Large Exp. Aquifer Program	Operating (1988)	Beaverton, OR	Industry, DOD, API, DOE, EPA	3 large, 2 medium-size in-ground steel tanks (outside)	Gasoline, solvents	Soil and groundwater	*In situ* mobilization transport and fate modeling	Air sparging, SVE, characterization tools (resistive tomography)	Modeling, transport and fate, mass balance

continued

Table 1.1 (continued) Summary of Planned and Operating Test Facilities in 1994–1995 (listed alphabetically by acronym)

	Stage of Development	Location	Funding	Facility Design	Contaminants of Interest	Media of Interest	Technology Categories To Be Tested	Technology Tests	Research Goals
RFETS Rocky Flats Enviro. Technology Site	Planning/design phase	Rocky Flats, CO	DOE	3 above-ground concrete "test beds" built-in sensor ports (in building)	Fuels, solvents, heavy metals, munitions, mixed wastes	Soil and groundwater	In situ and ex situ mobilization, destruction, containment	SVE, bio-venting reactive and flow barriers, electro-mobilization, soil separation, others	Promote development of remediation technologies beneficial to DOE
VEGAS Research Facility for Subsurface Remediation	Construction phase	Universität Stuttgart, Germany	Government, industry	Stainless steel tanks inside warehouse-like building (in building)	NAPL, PCE, PAH	Soil and groundwater	In situ extraction, decomposition, immobilization, bioremediation	Air sparging, surfactants, heat	Subsurface flow, mass transfer, chemical transformation
WES USAE Waterways Experiment Station	Operating	Vicksburg, MS	DOD	Corps of Engineers lab, work includes environmental tech. development and testing	DOD concerns: fuels, POL, solvents, munitions, heavy metals	Soil and groundwater	In situ destruction, characterization, mobilization, solidification	Based on DOD needs	Development of remediation technologies for Army cleanup/ compliance
WCS WCS Technology Development Center	Design phase; construction phase on landfill	Andrews County, TX	Venture capital, industry	Carbon steel in-ground tanks w/covers; other technology testing facilities (outside)	Fuels, solvents, heavy metals, munitions, mixed wastes	Soil and groundwater	In situ and ex situ contamination mobilization, stabilization, or destruction	Based on needs of DOD, DOE, and industry	Commercialization via venture capital or in-house funding
WATERLOO Waterloo Centre for Groundwater Research	Operating (1976)	Canadian Forces Base, Borden, Ontario	DOD, EPA, industry, others	Sheet piling to clay barrier, open top cell, funnel and gate barriers (outside)	Solvents, fuels, metals, creosote	Soil and groundwater	In situ reactive barriers, destruction, mobilization	Funnel and gate w/treatment cassettes, SVE, P&T, tracers, bioremediation, others	Transport and fate, mobilization, modeling, abandoned leaky boreholes
WURTSMITH National Center for Intergrated Biorem. R&D	Operating (1994)	Wurtsmith AFB, Oscoda, MI	SERDP/EPA Great Lakes HSRC	Releases without barriers, no cells (outside). plans to add tank	Fuels, solvents (fire training pit)	Soil and groundwater	In situ bioremediation	Surfactant-enhanced bioremed., co-metabolic bioventing enhanced anaerobic degradation of Cl-solvents	Aerobic/anaerobic bioremediation

Table 1.2 Design Comments from Advisors — ECRS Unit 1

Topic	Comments
Tank or cell size	An ideal but not practical cell or tank would be 60×30 ft deep $\times$ 100 ft long (15 ft saturated). For vadose zone technologies, the tank width should be 4 to 6 times the depth to groundwater. The cell length should be 200 times the groundwater velocity (ft/day); 0.5 ft/day = 100 ft. OGI's $60 \times 30 \times 20$ ft tank with 10-ft vadose zone (670 yd^3) is popular.
Modeling tank or cell size	The scale of the tank is dictated by number of experiments desired per technology. Larger tanks are limited to one experiment; they are slower, costlier, and more difficult to get mass balance. For design experiments, more runs in smaller tanks are needed. If running proof-of-concept tests, then one test in a large tank works. For some technologies, medium to small tanks are needed, where duplicate experiments can be run side-by-side or multiple experiments can be staged in sequence. Having two tanks packed with different soils (sands vs. clays) for testing the same technology would be helpful. Whether a tank or cell is used, it is still a model.
Test schedule	The larger the test cell or tank, the longer the test duration. Six months to a year is a minimum time for in-ground tests in cells or large tanks; 1 to 2 years is more typical. One year is a good time limit for medium-sized tanks.
Tank construction	To cut costs, prefabricated tanks could be used, such as fiberglass swimming pools.
Soil packing	To recreate swelling clays in tanks, buy Ca-smectite from distributor; mix with water and other soil components in commercial cement mixing equipment using paddles; line tank with geotextile for water infiltration around perimeter because it will not infiltrate well from the surface.
Clay tightness tests	Tests include permeability, Atterberg limits, grain size, mineralogy, suite of geotechnical tests. *In situ:* install piezometers for falling head test. Check water table level in sand aquifers above and below "sealing" clay unit. If water table in sand below the clay is lower than the water table in sand above the clay, the hydraulics will promote seepage of fluids downward through the clay.
Clay mineralogy	Clay used as base of test cells at Base Borden is glacial in origin, has no silt stringers, and is tighter than clays along the Gulf Coast, such as the Beaumont clay that does have silt stringers. The Beaumont would not be a suitable base for a test cell.
Other materials for test cell walls	Slurry walls are invariably permeable; do not use for ECRS cells. However, slurry walls with high density polyethylene liners are used for hazardous waste landfills and would provide better seal than sheet pilings.
Other designs besides tanks	Landfill design is the most secure release site (Title B or D landfills designated in 40 CFR regs 264, Title B hazardous waste; Title D municipal waste). Title B design uses two layers of geomembrane with heat-welded seams pressure tested for integrity. The lowest liner rests on 3 ft of clay with an engineered permeability of $\leq 10^{-7}$ cm/sec. A geonet material separates the liners. Upper liner is overlain by 2 ft of gravel and sand for protection and drainage. Sumps are located at each end, or beneath each bermed enclosure. A leak detection system can be installed between the liners.
Other test cells or tanks	Golder has a triaxial cell that GRI used in surfactant foam studies for remediation. The pressurized cell has been used to test fracture technology and to mimic injection delivery systems, among other things. It controls axial and radial stress and pore pressure.

Research on potential sites for the ECRS began with collection of information on industrial sites in Texas. Texas regulatory requirements were also studied in discussions with Nancy Worst, Director of Innovative Technologies, Waste Management Division, for the Texas Natural Resources Conservation Committee (TNRCC). The AATDF's DOD Advisory Committee was also approached to locate potential DOD sites. The DOD committee members were asked if their branch of the Service wanted to participate in the ECRS, and if so, to provide information on possible sites.

Four DOD organizations expressed interest in ECRS participation. They included the U.S. Army Engineer Waterways Experiment Station (WES), the Air Force Armstrong Laboratory, the Naval Facilities Engineering Command (NFEC), and the Naval Facilities Engineering Service Center (NFESC). The WES laboratory suggested locating the ECRS at WES, Vicksburg, MS. The Armstrong Laboratory suggested teaming with them to fund development of their Groundwater Remediation Field Laboratory, at Dover AFB, DE, and also provided data from their GRFL site search at other Air Force Bases, including Maxwell AFB, AL, Hill AFB, UT, and Charleston AFB, SC. NFEC responded with sites on two bases, Naval Air Station (NAS) North Island, San Diego, CA, and the NAS Mayport, FL, which were targeted by the Navy Environmental Leadership Program for cleanup. The NFESC offered test sites at the Naval Construction Battalion Center, National Test Site, at Port Hueneme and the Mare Island Ship Yard north of San Francisco, CA.

To evaluate these opportunities, an AATDF representative toured the WES laboratory facilities and discussed design options for a test release facility with Dr. John Cullinane and Mr. Norman Francinques. The design discussion included testing in portable tanks, such as the WES biotreatment cell that was operated in a soil-packed sludge container, with an adjacent shed that housed pumps and treatment equipment.

AATDF also met at Rice University with representatives of the Air Force Armstrong Laboratory, Major Mark Smith and Dr. Mark Noll, who were managing Air Force funding to develop a test release site, later named the Groundwater Remediation Field Laboratory (GRFL). At the time, the Armstrong Laboratory was involved in locating a site, completing a design and securing regulatory approval. The meeting agenda included an exchange of information regarding sites, schedules for development, focus of research, regulatory guidance, proposed designs, research clientele, teaming on facility development or on projects, and developing complementary research goals.

The NFEC and the NFESC provided site characterization data and related environmental reports to the AATDF for NAS North Island, CA, NAS Mayport, FL, the National Environmental Technology Test Site at Port Hueneme and the Mare Island Ship Yard, CA.

Information from the discussions with WES, the Armstrong Laboratory, the NFEC and NFESC, and the ECRS advisory committee helped the AATDF focus the strategy for ECRS development. Optimizing mass balance measurements during technology testing was a major objective for ECRS design and could best be met in a closed, containerized system, such as a pilot-scale tank. This consideration favored a portable size above-ground tank with a tightly sealable top and support equipment. With this design, a permanent in-ground facility with support staff would not be necessary, which precluded teaming with the Armstrong Laboratory on their sheet pile test cells (GRFL) or using sites offered by the NFEC or NFESC. The possibility still remained to locate a portable testing unit at the WES laboratory, with WES researchers providing funding for the research projects.

An ECRS consisting of portable testing units would serve the AATDF need for a unique market niche in technology testing, beyond the existing and planned test release facilities of the time. It also provided a more precise testing environment for controlled studies of technology efficiency or process mass balance. The design met many AATDF goals by providing:

- closed system, containerized design to improve mass balance determination and control environmental leaks
- intermediate-size test (pilot-scale) between bench-scale and full field-scale
- containerized testing unit at the researcher's own location for better monitoring
- standard engineering equipment for easy fabrication and maintenance
- flexible design to facilitate a range of soil packing conditions and technology tests
- no travel costs for research staff and their instrumentation

- reduced regulatory requirements, spill containment only and appropriate waste treatment
- low installation costs, only budget for shipping and maintenance costs
- no permanent staff
- tank size suited for projects of 1 to 1.5 years duration
- researchers responsible for disposal of waste soil and water, no liability to supplier (Rice University)

Information collected on portable tanks indicated that sludge containers could meet the design criteria for easy shipping, easy soil packing and removal, durable construction, tightly closed system, and availability. There were three general styles of sludge containers including roll-off or skid-mounted containers, trailers and rail-car gondolas. The roll-off and skid mounted tanks had standard 7-ft wide rectangular dimensions with volumes ranging from 10 to 40 yd^3 (lengths of 18 to 24 ft and depths of 3.5 to 7 ft). Floors were 3/16-in. or 1/4-in. steel with thinner walls reinforced by steel ribs. Floors had straight or rounded edges on the long sides. Interior of containers could have continuous welds with air pressure tests in gussets and corner. The tank interior could be coated with epoxy, and a variety of plastic liners were also available for any size container. Custom-fit lids were available with gaskets. Costs were relatively low, starting at approximately $6,000 for a small (10 yd^3) container with lid.

Trailer units had containers mounted on hydraulic lifts to operate as dump trucks. The trailer could be loaded with contaminated soil at a site and then driven to a test site and parked for testing. The trailer containers had less depth (3.5 ft) than the sludge containers and were longer (28 to 36 ft) with about the same width (7.5 ft). The cost was considerably higher at $19,000 to $23,000 per trailer. The rail-car gondola could have served as a significantly larger "portable tank." The interior of the container was approximately $7 \times 52 \times 10$ ft deep.

Draft Startup Plan and Draft Business Plan Outlines

On the questions of a site and design, the selection of a portable ECRS precluded the need for a permanent site and site manager. Regarding a partner for leveraging funding, AATDF decided to help support the first project at the Armstrong Laboratory GRFL test site and to develop a portable ECRS unit to be used at WES, in addition to a second ECRS unit that would travel to a researcher's location. A memorandum of understanding was signed with the Armstrong Laboratory and a CRADA was considered with the WES laboratory to operate the ECRS at the Vicksburg facilities. A draft business plan outline was prepared for development of the ECRS at WES, but the funding proved problematic because of their status as AATDF's grant administrator.

ECRS Project Selection and Initial Test Location

To facilitate systems tests and troubleshooting for the first unit, AATDF looked for a site in Houston for the first ECRS project. Potential sites included AATDF engineering partners and others with local offices and oil company research laboratories. Groundwater Services, Inc., Houston, TX, was contracted to provide warehouse space and engineering and design support to facilitate fabrication of the ECRS Unit 1, and Shell Development Company, Houston, TX, provided a secure site for assembly and operation of the unit and provided the first ECRS project.

For the second ECRS unit, AATDF looked for a project where the equipment could be shipped to the researcher's location to check the equipment and packing durability for long-haul shipping and to test the set up procedures. Protec, Dickinson, TX, was contracted to fabricate and ship the ECRS modules, and Arizona State University, Tempe, AZ, provided a secure site and an air sparging project for Unit 2.

CHAPTER 2

ECRS Engineering Design and Operation — Unit 1

2.1 DESIGN PARAMETERS

The ECRS should provide a cost-effective means of testing the efficiency of remediation technologies or processes at a fairly large pilot-scale under vadose or groundwater flow conditions. The guiding criteria for ECRS design were developed to meet this goal:

- The ECRS should consist of at least two portable units composed of modules that could be shipped to researchers' locations or remediation sites.
- The modular system should seal tightly to facilitate calculation of mass balances.
- Equipment must be fabricated using off-the-shelf components for easy assembly.
- The researcher should be able to pack the soil tank to simulate a variety of subsurface conditions.
- The equipment and instrumentation accompanying the ECRS should be able to accommodate a variety of contaminants.
- All instrumentation and equipment should be rugged enough to withstand normal shipping.
- All the modular equipment for each unit should ship on one flatbed trailer.

AATDF contracted with Groundwater Services, Inc. (GSI) to assist in preparing of the detailed design package for ECRS Unit 1. The design was prepared by Thomas Reeves and others at GSI with guidance from AATDF and Paul Johnson, ASU. GSI proposed a scope of work that included the following tasks:

Task 1: Project Startup
    Meet to define project goals and schedule
    Tour site at Shell, get health and safety requirements, and utility layout
    Review and revise the existing conceptual design
    Develop a detailed plan for implementation of the conceptual design
Task 2: Conceptual Design Review
    Evaluate the conceptual design with regard to the project goals and facilities at Shell
    Prepare and submit a preliminary design package consisting of a draft process flow diagram, results of process safety review, and identification of additional equipment and components
Task 3: Project Meeting
    Discuss results of conceptual design review, additional data needs, and assumptions for completion of detailed design

Task 4: Detailed Design Development
   Finalize design of ECRS Unit 1
   Develop a design package of engineering drawings and equipment specifications
      Process flow diagram
      Engineering flow diagram
      Electrical one-line diagram
      Equipment specifications
      Scope of work for project completion

The finalized engineering diagrams, including the process flow, engineering flow, and electrical one-line diagrams, are included in Section 2.3. The finalized equipment specifications are included in Section 2.4.

As part of Task 4, the scope of work for project completion included three additional tasks: project coordination, ECRS Unit 1 assembly, and ECRS system protective packaging. Project coordination involved a series of meetings with AATDF to review goals, specific design and equipment requirements, and the project completion schedule. ECRS Unit 1 assembly consisted of procurement and assembly of equipment and instrumentation for the soil tank, process equipment skid, instrumentation and data collection function, instrumentation building, interconnections between the modules, and toolbox for routine system operations and maintenance.

Overview of Equipment and Costs

The assembled unit consisted of four modules, including an instrumentation building, process equipment skid, soil tank, and reservoirs. AATDF contracted with Bebco for fabrication of the insulated instrumentation building with climate control, bench space for analytical and computer equipment, an exterior bottled gas rack, stainless steel piping for GC gases, electrical outlets, and storage space. The process equipment skid was fabricated by GSI and was equipped with a compressor, blower, groundwater pumps, water and air filters, piping, an electrical panel, a gauge panel, and a data recording/display unit for real-time acquisition and monitoring of system parameters and controls. For the soil tank, AATDF purchased a standard 27-yd^3 rectangular, reinforced steel sludge container from Galbreath Inc., Mansfield, TX, with interior epoxy coating and a water tightness test certification. Galbreath Inc. made the following modifications to the sludge container: addition of 12 2-in. OD pass-thru pipes with threads and caps inserted in a row 2 in. below the top cap on one wall, and in three groups of four holes; eight 2 in. OD pass-thru pipes with threads and caps in all four corners of both long walls; two horizontal rows of Unistrut welded along the outside of the two long walls; four vertical pieces of u-channel welded inside the two long walls; and the ladder moved from the side to front wall.

The tank was retrofitted by GSI with flanges, tubing nozzles, valves, tee-strainers, and sight gauges. AATDF purchased the flexible tank top from Mesa Rubber Co., Houston, TX. At least three cost bids were solicited before purchase of equipment. When a sole source purchase was necessary, it was documented with the appropriate paperwork. Table 2.1 provides a representative list of the equipment and suppliers for ECRS Unit 1.

The design, assembly, and testing of ECRS Unit 1 was documented in AATDF report TR-96-2 (AATDF 1996). Portions of this report are discussed in the following sections.

2.2 UNIT 1 MODULES OVERVIEW

The ECRS was designed to be a modular, portable, and durable research tool for testing remediation technologies in a contained, yet realistic, setting. The system was built at a pilot scale using full-size standard equipment to provide cost-effective demonstrations at a scale between laboratory

Table 2.1 Summary of Major Equipment (1995–1996 Prices) — ECRS Unit 1

Description	Vendor	No.	Total Cost
Tank with modifications, but w/o lid tank model OS1872-6	Galbreath, Inc. Mansfield, TX	1	$8,931
Geotextile (1st cover for tank) HDPE 60 mm	GSE Lining Technology Inc. Gundle Houston, TX	1	$500
Tank top (2nd top for tank) material MESA6036.35FCA	Mesa Rubber Co., Houston, TX	1	$886
Steel frame, fabric cover, sun/rain shelter	Hansen Weather Port Gunnison, CO	1	$5,900
Steel shed, skid base, AC-heater, electrical hookups, transformer, light, wiring, counterspace	Bebco Industries Texas City, TX	1	$10,000
Soil vacuum extraction pump system and air compressor system, two skid-mounted units SpargePro, SitePro	ORS Greenville, NH	1 SVE 1 sparge	$20,274
Continuous FID/PID vapor analyzer VIG Model 20	VIG Industries Chino, CA	1	$16,429
Helium detector (tracer gas)	Mark Products, Inc. Sunnyvale, CA	1	$5,300
Vortex flowmeter Swingwirl II DMV 6336Z	Endress & Hauser Houston, TX	4	$1,678
Pressure transmitter Cerabar S PMC731 0521F6H11N1	Endress & Hauser Houston, TX	6	$823
Pipe mounting kit for Cerabar S, model 7039	Endress & Hauser	6	$190
Air injection flow meters (0–20 scfm) with valves Dwyer VFC-121-SSV	Texas Process Equipment Houston, TX	5	$435
Vapor extraction flow meters Dwyer VFC-122-SSV	Texas Process Equipment Houston, TX	5	$435
Groundwater recirculation flow meter (0.1–1 gpm) Dwyer RM-141-SSV	Texas Process Equipment Houston, TX	2	$140
Groundwater recirculation flow meter (0.02–0.3 L/min) Dwyer RM-34-SSV	Texas Process Equipment Houston, TX	2	$140
Air injection pressure gauges (0–500 in. H_2O) Dwyer Capsuhelic 4500	Texas Process Equipment Houston, TX	5	$350
Vapor extraction vacuum gauges (0–200 in. H_2O) Dwyer Capsuhelic 4200	Texas Process Equipment Houston, TX	5	$350
Water level gauges (0–500 in. H_2O) Dwyer Capsuhelic 4616 w/bleed valves	Texas Process Equipment Houston, TX	2	$140
Differential water level gauge (0–25 in. H_2O) Dwyer Capsuhelic 4025 w/bleed valves	Texas Process Equipment Houston, TX	2	$140
Groundwater recirculating pump (0.1–1 gpm@10ft head)	Texas Process Equipment Houston, TX	2	$3,252
Groundwater drain/fill pump (0–20 gpm@10ft head)	Texas Process Equipment Houston, TX	2	$1,984
PVC threaded gate valves	Texas Process Equipment Houston, TX	10	$170
Stainless steel fittings, valves, and tubing	Whitson & Co. division of Rawson & Co. Houston, TX		$4,495
Process equipment skid fabrication, plus Dwyer L6EPB-S-s-3-0 float switches filter housings, mountings, and filters Differential pressure gauges	Eggelhof Inc. Houston, TX		$7,681
Masterflex Portable Pump (H-07570-10) Batteries (H-07578-60) Heads (H-07018-21)	Barnant Co. Barrington, IL	2 2 2	$2,070

continued

Table 2.1 (continued) Summary of Major Equipment (1995–1996 Prices) — ECRS Unit 1

Description	Vendor	No.	Total Cost
Thomas vapor sampling pump Thomas 917CA18 & TFEL pumps	Wistech Controls Wilson Co. Dallas, TX		$600
2 1200-gal vertical heavy duty tanks and bulkhead fittings	Chem-Tainer Houston, TX		$1,730
55-gal drums of carbon for vapor and water treatment	Calgon Carbon Corp. Houston, TX	2 air 2 H$_2$O	$2,234
Basic manual tool kit, drive point device	Geoprobe Salina, KS	1	$1,125
Data logger, multiplexer	Campbell Scientific Logan, UT	1	$3,391
Portable notebook computer	CompUSA Houston, TX	1	$3,599
		Total	$105,372

bench-scale and full-scale field implementation. The equipment can simulate a range of operating conditions, contaminant distributions, and geologic settings. It can be configured to simulate a variety of physical site conditions, such as groundwater or vadose zone environments or subaquatic sediments, and to support biological growth and testing on both a microbial scale and for plant propagation. The recirculating water and SVE systems can be used to add chemical amendments or tracers to enhance or to characterize physical, chemical, or biochemical remediation activities.

The ECRS consists of four well-engineered modules that include the instrumented soil tank, process equipment skid, instrumentation building, and water reservoirs. Each module is transportable by forklift, and the entire system is packed onto one low-boy trailer for shipment to a researcher's location. Having the ECRS on site allows researchers to supervise all phases of their experiments using their own instrumentation and staff. This eliminates the researcher's need for travel to a remote field site.

2.2.1   Process Equipment Skid

The process equipment skid for Unit 1 is shown in Photo 2.1 and Appendix A, Figure A3.2. It is constructed of carbon steel with welded brackets and supports, and is equipped with a compressor, blower, air and water regulation valves, two water pumps, air and water filters, and a control panel. The compressor that pumps ambient air into the tank is a 5-hp centrifugal air pump capable of

Photo 2.1   Process equipment skid — Unit 1.

producing 35 cubic feet per minute (cfm) at 15 pounds per square inch gauge pressure (psi). A pressure relief valve is set to 15 psi on the exhaust side of the compressor.

The blower that pulls air from the tank is a 5-hp positive displacement rotary lobe blower equipped with input and output silencers. The blower can produce a total air flow of 100 cfm at a vacuum of 14-in. in Hg. To prevent buildup of pressure in the tank and inflation of the top, a differential pressure-reducing regulator and a back pressure regulator are set to maintain a positive pressure differential between the air flows. The blower circuit includes an inline moisture separator with automatic drain, an air filter, and a vacuum relief valve.

Water is circulated through the soil in the tank using a pair of duplexed 0.5-hp magnetically coupled gear pumps. The pumps are capable of producing 1 gal/min each for a total flow of 2 gal/min. Water from the tank passes through a series of three cartridge filters (100, 50, 50 μm) to remove fines before it enters the pumps.

2.2.2 Instrumentation Building

The instrumentation building for Unit 1 is shown in Photo 2.2 and Appendix A, Figures A5.1 and A5.2. It is constructed on a steel channel skid with forklift channels to make it easily transportable. The building itself is constructed of galvanized 18-gauge steel and glass fiber insulation and has the outer dimensions of $7 \times 6 \times 9$ ft tall. It is a well-lighted, well-insulated structure with climate control (an air conditioner and heater) and positive pressure to protect the instruments and researchers from extreme temperatures and vapor buildup. The building houses system controls, analytical instruments, and power. The system controls consist of a data logger (Campbell Scientific CR-10) and a 486 PC-based computer. The analytical instruments include a helium detector (Mark Products 9823) and a total hydrocarbon analyzer (VIG Industries Model 20) that receives samples through a heated line from the soil tank. Accessories include hookups and piping for gas cylinders, counters on two walls for instrumentation, and an overhead cable tray for wiring. The power supply can come from the building's 3-phase, 480 volt transformer or from a portable generator supplied by the researcher.

Photo 2.2   Instrumentation building — Unit 1.

2.2.3 Soil Tank

The ECRS soil tank, shown in Photo 2.3 and Appendix A, Figure A3.1, is a 27-yd^3 rectangular container with dimensions $18 \times 7 \times 6$ ft deep, which was fabricated by Galbreath Inc. (Model

Photo 2.3   ECRS Unit 1 tank setup at Shell WTC, Houston, TX.

OS1872). It has heavy-duty reinforced steel construction (3/16-in. gauge floor, 12-gauge sides, side stakes on 36-in. centers, all continuous welds) with a gasketed rear door (1-in. thick, 2.5-in. wide rubber gasket) for easy soil disposal after testing. Vertical u-channel tracks are welded on the inside walls at each end of the tank to hold temporary plywood partitions during soil packing. These partitions facilitate placement of a selected sand pack over horizontal well screens in the water intake and extraction zones at opposite ends of the tank.

The interior of the tank is coated with black polyamid epoxy to prevent corrosion and to improve water tightness. The polyamid epoxy coating on the Unit 1 tank is compatible with fuels and fuel oil contaminants and can be replaced with other coatings resistant to selected contaminants such as solvents. The entire tank is leak tested before delivery.

The tank is equipped with 15 nozzles (2-in. OD pass-thru pipes) that can be configured using a variety of connections to inject or extract air or water, or to sample vapors or liquid. The outside front of the tank has two instrument wiring harnesses bolted to a horizontal Unistrut channel that is welded to the outer tank wall. The instrumentation on these harnesses includes two hydrostatic pressure transmitters, sparge and vent manifolds with flow meters, pressure transmitters, and two sample port manifolds with dozens of sample ports.

The top of the soil tank is sealed with a soft fabric cover. Both soft and hard tops were considered for the tank. The soft top was selected for a number of reasons. It is lightweight and easy to remove and replace by hand instead of with a crane. Manways can be added to the top for access to the soil pack. The soft top reduces any potential explosive hazard associated with the enclosed tank.

The original top for Unit 1 was composed of 60-mm thick Gundle high-density polyethylene (HDPE) landfill liner without manways. The flat edge of the HDPE was mechanically sealed against the 5-in. wide lip of the tank with a 0.5-in. neoprene gasket and silicon caulk, and secured with angle iron and c-clamps. To improve the mechanical seal, the tank lip had to be prepared for placement of the gasket and caulk; the metal surface was smoothed with a grinder and irregularities were filled with epoxy. The surface of the HDPE was covered with a tarp to reduce incident sunlight.

A second soft top has been fabricated from two-ply urethane (Mesa 6036.35 FCA, one-ply ester-based and one-ply ether-based) to provide enhanced long-term durability, low sorption, and low permeability to fuel contaminants. It has two carbon steel manways with Teflon bolts to provide access to the soil and instrumentation after the top is sealed.

2.2.4 Reservoirs and Treatment Equipment

The water reservoirs for Unit 1 are two 1200-gal white, heavy-duty polyethylene tanks (Photo 2.4 and Appendix A, Figure A3.3). Each tank is 4.6 ft tall and 7.1 ft in diameter. Each reservoir has two drain and fill nozzles. The fill connection is at the top of the tank, and the drain connection is 3 in. from the base. The tanks can act as water reservoirs, emergency holding tanks for draining the ECRS tank, and as mixing tanks for tracers and other chemical amendments. The reservoirs have been covered with black plastic to inhibit algal growth, which was promoted by summer sunlight and heat after the ECRS unit was set up at Shell Development Company.

The choice of appropriate water and vapor treatment equipment for the air and water waste streams from ECRS units is the responsibility of the researcher. The treatment systems selected for use with Unit 1 during the first air sparging projects with BTEX components were 55-gal drums of granular activated carbon.

Photo 2.4   Unit 1 modules.

2.3 UNIT 1 ENGINEERING DESIGN

The engineering design for ECRS Unit 1 was provided by GSI and was critiqued and modified by the AATDF staff and advisors. Eleven design figures were prepared for the three-volume Operation and Maintenance Manual for ECRS Unit 1 (AATDF 1997a). They are listed below and included in Appendix A:

Figure A1.1   Process/engineering flow diagram, general notes and symbols
Figure A1.2   Process/engineering flow diagram, instrumentation notes and symbols
Figure A1.3   Electrical diagram, general notes and symbols
Figure A2.0   Process flow diagram
Figure A3.1   Engineering flow diagram, ECRS soil tank
Figure A3.2   Engineering flow diagram, sparging/SVE groundwater package
Figure A3.3   Engineering flow diagram, ancillary equipment
Figure A4.1   Electrical diagram, power distribution
Figure A4.2   Electrical diagram, data collection wiring harness
Figure A5.1   Detail diagram, instrumentation building exterior
Figure A5.2   Detail diagram, instrumentation building interior

2.4 DESCRIPTION OF SYSTEM FUNCTIONS

Primary system components are described below and in Tables 2.2 to 2.5. Equipment and instrumentation specifications are outlined in Tables 2.2 to 2.4, which also provide lists of equipment numbers used on the design diagrams for Unit 1.

- ECRS Soil Tank — The soil tank and connections are configured to facilitate various schemes for sampling, injection, and extraction of water and/or air. This 27-yd^3 rectangular tank is equipped with nozzles and is filled with porous, permeable media. Prior to the start of each experiment, the permeable media within the soil tank can be spiked with a known type and amount of contaminant. During the project, other chemical amendments that are compatible with the lining on the tank and the composition of the connectors and hoses can be added to the soil, air stream, or circulating water.
- *Soil Vapor Extraction System* — Contaminated air within the soil matrix can be removed with the soil vapor extraction (SVE) blower. The SVE assembly consists of an inline moisture separator, particulate air filter, vacuum relief valve, and a 5-hp rotary lobe blower equipped with inlet and outlet silencers. The schedule of instrumentation and electrical components for all these systems is included in Table 2.3, and the schedule of valves is listed in Table 2.4.
- *Air Injection System* — A 5-hp compressor allows for air injection into the soil within the ECRS tank. Air can be introduced into the soil via a wellpoint constructed of a casing fitted with a well screen or via other devices designed by researchers for specific projects.
- *Water Recirculation* — Water is recirculated through the soil tank using duplexed 0.5-hp gear pumps. Prior to passing through the pumps, water exiting the soil tank is filtered through coarse and fine spiral wound cartridges to remove soil particles with diameters greater than 50 μm.
- *Water and Air Treatment* — Air and water waste streams extracted from the soil tank may require treatment to remove any contaminants prior to discharge into the atmosphere or an appropriate wastewater collection system. For the first system test of ECRS Unit 1, separate granular activated carbon (GAC) units were used to remove dissolved organic compounds from water and air extracted from the soil tank before being recirculated through the soil media. Single-use GAC units were contained in 55-gal drums obtained from Calgon. During normal ECRS operation, researchers are responsible for providing their own treatment systems.
- *System Controls* — System control features were revised in the fall of 1997. They originally included a control panel mounted on the SVE and an air injection equipment skid to provide for system control and shutdown in the event of an emergency. The SVE and air injection systems were hardwired to allow operation of the system only when both were functional.
- *Data Acquisition* — A data logger installed in the instrumentation building receives and records signals from instruments and sensors mounted in the soil pack or on the soil tank and piping. Instruments such as pressure sensors, flow meters, and temperature sensors provide a direct reading of conditions within the soil tank and associated equipment. The data logging software can be customized for real-time data acquisition of parameters required for each experiment. The data logger components are listed in Table 2.5.

The ECRS unit has been designed and constructed to provide maximum flexibility for implementing various experimental configurations. In order to support the ECRS equipment, certain site requirements need to be considered. These include the following:

- *Workspace* — A minimum area of 25 × 50 ft is required to allow adequate clearance between the components. A means of securing the site is also necessary for personnel safety and to prevent equipment theft or damage. Depending upon the chemicals and their concentrations involved in the ECRS tests, secondary containment or access to a chemical sewer may be required.
- *Utilities* — The preferred electrical connection for the ECRS unit power supply is a 4-wire, 480 volt, 50 amp service with 3-phase conductors and a grounding conductor. Alternatively, a 208 volt,

TABLE 2.2 Major Equipment Specifications — Unit 1

Number	Name	Service	Media Data		Description		Manufacturer
3-B-02	Soil Sparging Compressor	Ambient air	Viscosity: Density: Temp: Abrasives: Solids:	0.018 cP 1 lb/ft³ Ambient Light <1%	Type: Power:	Centrifugal 5 hp, 230 V,3Ø	Gast Model 6066
2-B-01	SVE Blower	Recovered soil vapor	Viscosity: Density: Temp: Abrasives: Solids:	0.018 cP 1 lb/ft³ 170°F Light <1%	Type: Inlet pressure: Temp rise: Pressure rise: Power:	Positive displacement, rotary lobe 14 in. Hg (max) @ 3600 rpm 170°F 12 psi 5 hp, 230 V,3Ø	R&M Associates Model 33-URAI
3-F-05	Soil sparging compressor pre-filter	Ambient air	Solids size:	—	Felt		ORS Environmental Systems Model AD 750
2-F-04	SVE blower pre-filter	Recovered soil vapor	Solids size:	1 µm	Pleated paper		Stoddard, Inc. Model FGS-2
1-F-01	Water recirc. pump coarse pre-filter	Recovered water	Solids size:	100 µm	Type: Pressure (max): Temp (max):	Spiral wound 300 psig 200°F	CT101A (housing) 47310-01 (filter) CUNO Meridien, CT
1-F-02 1-F-03	Water recirc. pump fine pre-filter	Recovered water	Solids size:	50 µm	Type: Pressure (max): Temp (max):	Spiral wound 300 psig 200°F	CT101A (housing) 47310-01 (filter) CUNO Meridien, CT
1-GAC-01 1-GAC-02	Granular activated carbon	Recovered water					Calgon Corp.
2-GAC-03 2-GAC-04	Granular activated carbon	Recovered soil vapor					Calgon Corp.
1-P-01 1-P-02	Water recirc. pump	Recovered water	Viscosity: Sp. Gr.: Temp: Abrasives: Solids:	1cP 1 Ambient Light <1%	Type: Inlet Pressure: Max. Pressure: Power:	Magnetically coupled, gear 0 psi 40 psi 0.5 HP	
2-SEP-01	Moisture separator for SVE blower	Recovered soil vapor	Capacity:	15 gal	High efficiency, cyclonic separator		ORS Environmental Systems
1-T-01	ECRS tank	Water	Characteristics of soil and contaminants will vary per parameters of experiment		Type: Capacity: Standards:	Steel rolloff box 27 yd³ ANSI Z245.1 and Z245.2 safety standards	Gilbreath, Winamack, IN Job No. 18170 Model No. OW1872-6
1-T-02 1-T-03	Water reservoir tanks	Potable water	1000 gal		HDPE		

Table 2.3 Schedule of Instrumentation and Electrical Components — Unit 1

Service	Area	Item No.	Description
Water Circulation	ECRS Tank Connections	1-PIT-221	Pressure Transmitter
		1-LSHH-222	High Level Alarm
		1-PIT-223	Pressure Transmitter
		1-LSHH-224	High Level Alarm
	Filter Assembly	1-PDG-261	Differential Pressure Gauge
	Recirculation Pumps	1-PG-281	Pressure Gauge
		1-FE-422	Flow Indicator
	Water Treatment Manifold	1-PDG-421	Differential Pressure Gauge
Air Circulation	SVE Effluent from ECRS Tank	2-PIT-141	Pressure Transmitter
		2-FG-142	Flow Meter
		2-FIT-143	Flow Transmitter
		2-PIT-144	Pressure Transmitter
		2-FG-145	Flow Meter
		2-FIT-146	Flow Transmitter
	SVE Blower Assembly	2-LSHH-181	High Level Alarm
		2-FG-182	Flow Indicator
		2-IT-183	Current Transmitter
		2-HS-184	Hand Switch
		2-TS-185	Temperature Switch
		2-TS-186	Temperature Switch
	Soil Sparging Influent to ECRS Tank	3-FIT-121	Vortex Shedding Flow Meter
		3-PIT-122	Pressure Transmitter
		3-FIT-125	Vortex Shedding Flow Meter
		3-PIT-124	Pressure Transmitter
	Soil Sparging Compressor Assembly	3-HS-161	Hand Switch
		3-IT-162	Current Transmitter
		3-PG-163	Pressure Gauge

TABLE 2.4 Valve Specifications — Unit 1

Service	Area	Item No.	Description
Water Circulation	ECRS Tank Connections	1-HV-181	Control Valve
		1-HV-182	Control Valve
	Influent/Effluent Manifold	1-HV-231	Control Valve
		1-HV-232	Control Valve
		1-HV-233	Control Valve
		1-HV-234	Control Valve
		1-HV-235	Control Valve
		1-HV-236	Control Valve
		1-HV-237	Control Valve
	Filter Assembly	1-HV-251	Control Valve
		1-HV-252	Control Valve
		1-HV-253	Control Valve
		1-HV-254	Control Valve
		1-HV-255	Control Valve
		1-HV-256	Control Valve
		1-HV-257	Control Valve
		1-HV-258	Control Valve
		1-HV-259	Control Valve
		1-YS-260	Y-Strainer
	Circulation Pumps	1-HV-271	Control Valve
		1-HV-272	Control Valve
		1-HV-273	Control Valve
		1-HV-274	Control Valve
		1-HV-275	Control Valve

TABLE 2.4 (continued) Valve Specifications — Unit 1

Service	Area	Item No.	Description
Water Circulation	Water Treatment Manifold	1-HV-411	Control Valve
		1-HV-412	Control Valve
		1-HV-413	Control Valve
		1-HV-414	Control Valve
		1-HV-415	Control Valve
		1-HV-416	Control Valve
	Water Reservoir Tanks	1-HV-431	Control Valve
		1-HV-432	Control Valve
		1-HV-433	Control Valve
		1-HV-434	Control Valve
	Liquid Phase Carbon	1-HV-451	Control Valve
		1-HV-452	Control Valve
		1-HV-453	Control Valve
Air Circulation	SVE Effluent from ECRS Soil Tank	2-HV-140	Sample Port
		2-HV-141	Control Valve
		2-HV-142	Sample Port
		2-HV-143	Control Valve
		2-HV-144	Sample Port
		2-HV-145	Control Valve
		2-HV-146	Sample Port
		2-HV-147	Control Valve
		2-HV-148	Isolation Valve
		2-HV-149	Bleed Valve
	SVE Blower Assembly	2-PRV-171	Pressure Relief Valve
		2-PRV-172	Manual Dilution Valve
	Vapor Phase Carbon	2-HV-311	Control Valve
		2-HV-312	Control Valve
		2-HV-313	Control Valve
	Soil Sparging Influent to ECRS Soil Tank	3-HV-100	Isolation Valve
		3-HV-101	Control Valve
		3-HV-102	Sample Port
		3-HV-103	Control Valve
		3-HV-104	Sample/Injection Port
		3-HV-105	Control Valve
		3-HV-106	Sample Port
		3-HV-107	Control Valve
		3-HV-108	Sample Port
		3-HV-109	Bleed Valve
	Soil Sparging Compressor Assembly	3-HV-151	Throttling Valve
		3-CKV-152	Check Valve
		3-PRV-153	Pressure Relief Valve
		3-HV-154	Sample/Injection Valve
		3-DPRV-155	Differential Pressure Regulating Valve
		3-PRV-156	Pressure Regulating Valve

TABLE 2.5 Data Logger Components — Unit 1

Name	Description		Model
Rechargeable sealed lead-acid battery	6V 8Ahr		Model PE6VS, Japan Storage Battery Co., Ltd.
DC power supply	Input:	100/120/220/240 VAC ±10% 47–440 Hz	FDBB-200W-I, Condor, Inc.
	Output:	+5VDC@10.5 A/12 pk OVP at 6.2 ± 0.4V −12VDC@1.5 A/2.0 pk +24VDC@7.5 A/8.5 pk	

continued

TABLE 2.5 (continued) Data Logger Components — Unit 1

Name	Description		Model
Class 2 transformer	Input: Output:	120 VAC 60 Hz 30W 18VAC 1.11A	830A0064-03, Tamura
Power supply	12V charging regulator		PS12, Campbell Scientific, Inc.
Optically isolated RS 232 interface	Input: Output:	Data logger Terminal/printer	SC32A, Campbell Scientific, Inc.
Peripherial to RS232 interface	Input: Output:	Peripheral RS232	SC532, Campbell Scientific, Inc.
Keyboard display	Input:	Serial input/output	CR10KD, Campbell Scientific, Inc.
Relay multiplexer			AM416, Campbell Scientific, Inc.
Measurement and control module			CR10, Campbell Scientific, Inc.
Storage model			SM716, Campbell Scientific, Inc.
Control panel enclosure	16 × 10 × 18 in.		Vynco

100 amp service could be used, provided a 3-phase, 5-wire system were available with 3-phase conductors, a grounded neutral, and a grounding conductor.

- *Health and Safety* — A commitment is needed on the part of the principal investigator to provide controls for the health and safety of all personnel involved in the installation and operation of the ECRS. This will include familiarity with applicable workplace safety regulations, as well as personnel training and site inspections (Appendix B).

2.5 PROCESS CONTROL

Appendix A, Figures A3.1, A3.2, and A3.3, respectively, show engineering flow diagrams; the air injection, SVE, and groundwater systems of the ECRS soil tank; and the ancillary equipment. A three-part nomenclature (e.g., 1-T-01) for equipment and instrumentation is employed on the figures and in the description below. The first number in the three-part label refers to the subsystem (i.e., 1 for water, 2 for SVE, and 3 for air injection). Standard equipment and instrumentation references are used for the second part of the identifier (Appendix A, Figure A1.2). The last part of the label is a unique two- or three-digit numerical designation for the system component. Lists of equipment and instrumentation numbers for Unit 1 are included in Tables 2.2 to 2.4.

ECRS Soil Tank

The soil tank (1-T-01) is a modified 27-yd^3 rectangular sludge container equipped with nozzles that may be connected in various configurations for sampling, injection, and extraction of vapor or water. For the purpose of experimental testing of remediation technologies or processes, the tank can be packed with soil or other porous, permeable media and equipped with sensors and samplers.

Water Recirculation

Water is recirculated through the soil within the soil tank (1-T-01) using a pair of duplexed 0.5-hp magnetically coupled gear pumps (1-P-01, 02). The water recirculation pumps are capable of producing 1 gpm each for a total flow of 2 gpm. Water for filling the ECRS soil tank is stored in two 1,200-gal polyethylene tanks with 1-in. cam-lock fittings (1-T-02, 03).

Filtration

To remove fine particles suspended in the wastewater stream produced from the soil tank, the water first passes through a coarse (100 μm) cartridge filter (1-F-01). After coarse filtration, water recovered from the soil tank passes through two fine (50 μm) cartridge filters connected in series (1-F-02, 03). The upstream fines filter (1-F-02) needs to be replaced when the pressure drop exceeds 10 psi as indicated on a differential pressure gauge (1-PDG-261). If the pressure drop returns to an acceptable level, no further action is necessary. If a pressure drop of 10 psi or more is still indicated by the differential pressure gauge (1-PDG-261), then the down stream fines filter (1-F-03) should be changed. Should the pressure drop not return to an acceptable level, then the coarse filter (1-F-01) should be changed.

SVE System

Air contained within the soil pore spaces in the ECRS tank (1-T-01) is removed by the SVE blower (2-B-01). The SVE assembly includes an inline moisture separator with automatic drain (2-SEP-01), an air filter (2-F-04), a vacuum relief valve (2-PRV-171), and a 5-hp positive displacement rotary lobe blower (2-B-01) equipped with input and output silencers. The SVE assembly is capable of producing a total air flow of 100 cfm at a vacuum of 14 in. Hg.

Air Injection System

Ambient air is pumped into the ECRS soil tank (1-T-01) by a compressor (3-B-02), a 5-hp centrifugal air pump capable of producing 35 cfm at 15 psi. On the exhaust side of the compressor (3-B-02), a pressure relief valve (3-PRV-153) is set to 15 psi. To prevent pressure buildup inside the soil tank when the top has been sealed, a differential pressure-reducing regulator (3-DPRV-155) and a back pressure regulator (3-PRV-156) are set to maintain a positive pressure differential between the SVE and air injection flows.

Water and Air Treatment

Options for treatment of wastewater and air are discussed in a previous section. Treatment of wastewater and air from the soil pack spiked with BTEX, using GAC, follows:

- Water treatment — Water recovered from the soil tank (1-T-01) is pumped through two liquid phase GAC drums (1-GAC-01, 02) to remove dissolved BTEX contaminants before being recirculated to the soil tank. New drums of GAC should be obtained from the vendor when analytical testing indicates that effluent water no longer meets the requirements for reuse or disposal.
- Air treatment — Two vapor phase GAC drums (2-GAC-03, 04) remove BTEX contaminants from the air extracted from the soil tank. Treated air is directly discharged to the atmosphere. New GAC should be obtained from the vendor when effluent air no longer meets required characteristics for discharge.

Piping

- Water piping — The soil tank (1-T-01) has two water connections, each constructed of a threaded nipple welded through a hole in the side of the tank. The threaded nipple is connected to a threaded fitting in which a 1-in. flange has been welded and connected to a ball valve (1-H-181, 182). Water enters the tank from the water circulation pump via line GS-14-1 by the groundwater recirculation pump(s) (1-P-01, 02). For maximum flexibility, water collection and discharge piping is constructed of 1-in. diameter hose that connects the soil tank to the process equipment skid and of 1.5-in. diameter stainless steel tubing on the skid. Connections are made by flanges or swaged fittings.
- Air connections — The soil tank is currently configured with two manifolds. One is an influent for air injection and the other is an effluent for SVE (Appendix A, Figure A3.1). For maximum flexibility, air collection and discharge piping is constructed of 1- and 2-in. diameter hose from the soil tank to the process equipment skid and to the vapor-phase GAC units.

Secondary Containment

ECRS Unit 1 is designed for experiments using potentially hazardous and/or regulated chemicals. Therefore, each site will need to consider secondary containment to collect any releases of affected water and chemicals or to direct them into an appropriate chemical-process sewer.

System Controls

A control panel mounted on the process equipment skid provides for automatic shutdown of the system should an alarm condition occur. Interlocks providing for system shutdown include the following (Appendix A, Figure A3.2):

- Soil tank high water — The soil tank is protected by high water-level switches on the influent and effluent water zones (1-LSHH-222, 224). Should the water level in the tank rise above the high level switches, power to the groundwater recirculation pumps (1-P-01, 02) will be cut off.
- SVE moisture separator high water — The moisture separator (2-SEP-01) on the influent line to the SVE blower is protected by a high water-level shutoff switch (2-LSHH-181). If the water level within the moisture separator is exceeded, power will be cut off to all air and water circulation systems.
- SVE/air injection high pressure — To prevent damage to the soil tank fabric cover from overpressuring the tank, connections have been hardwired to the motor controls for the SVE blower (2-B-01) and air injection compressor (3-B-02) so the system operates only when both are running.

Data Acquisition and Recording

The instrumentation building, located adjacent to the process equipment skid, houses data logging hardware, analytical instruments, power connections, and climate control. Depending on the site configuration, system conditions can be monitored continuously through pressure gauges, temperature sensors, and flow meters on air and water lines connected to the soil tank.

Shielded cables convey signals from instruments located within the soil tank and piping to the instrumentation building. Inside the building, signals from the system sensors are received by a 32-channel CR-10 data logger. The data logger needs to be reprogrammed at each site to correctly read and store information from the input/output data points indicated on Appendix A, Figures A3.1 and A3.2. The site configuration during the system test provided for acquisition of water and air process variables, as follows:

- Groundwater data — Recirculated water flow rate is measured using an electromagnetic flow meter (1-FIT-422) located on the process equipment skid. In order to determine the differential piezometric pressure within the tank, a pressure-indicating transmitter is installed at each end of the tank (1-PIT-221, 223).
- SVE data — Two vortex shedding flow meters (2-FIT-143, 146) on the SVE manifold mounted to the soil tank provide information locally and to the data logger regarding SVE air flow rates out of the soil tank. Pressure in these lines is monitored by two pressure sensors (2-PIT-141, 144).
- Air injection data — Two vortex shedding flow meters (3-FIT-121, 125) on the air injection manifold mounted to the soil tank provide information locally and to the data logger regarding air flow rates into the soil tank. Pressure in these lines is monitored by two pressure sensors (3-PIT-122, 124).

2.6 HEALTH AND SAFETY

Operation and maintenance of the ECRS should be conducted in accordance with the site-specific project health and safety plan. Appendix B provides a generic format that can be used in developing the site-specific health and safety plan. All personnel working with the ECRS should read and be familiar with the health and safety plan. The following sections provide additional information to supplement that included in Appendix B. On-site responders to fire/spill conditions must be apprised of site hazards and chemicals.

Site Hazards

- Night Operation — Work areas will need adequate lighting for researchers or other personnel to see to work and identify hazards. Applicable OSHA standards for lighting apply (29 CFR 1910.120m).
- Electrical Power — The circuit breaker, transformer, motor starters, and control boxes are significant electrical hazards and should be serviced only by authorized personnel. The power distribution for the ECRS soil tank is equipped with ground-fault interrupt circuit breakers. The electrical hazard class around the ECRS soil tank will be site- and project-specific. All electrical equipment meets the Class 2, Division 1 specifications. Applicable OSHA standards for electrical equipment apply (29 CFR 1926 Subpart K).
- Rotating Equipment — Rotating equipment consists of the compressor, blower and pumps, and the process equipment skid. Guards are installed over the drive couplings on the compressor and pumps and on the belt of the blower. These guards protect against incidental contact and injury from rotating parts, and the equipment should not be operated without these guards in place. When the guards are removed for equipment maintenance, the personnel performing the maintenance must replace them prior to operation of the equipment.
- Noise — Both the air compressor and extraction blower on the process equipment skid produce elevated noise levels. Anywhere that noise levels exceed 85 db, OSHA regulations specify that personnel should wear hearing protection. That level can be exceeded near the compressor and blower units. If the site configuration permits, the process equipment skid can be isolated from normal personnel workspaces by distance or intervening equipment.
- Chemical Hazards — Various types and concentrations of chemicals can be used during testing in the ECRS soil tank, analytical work in the instrumentation building, or cleaning of equipment or instrumentation. Those solutions or gases could include contaminants such as hydrocarbon fuels, petroleum oils and lubricants (POL), or solvents, chemical amendments for treatment, tracers, pressurized gases, analytical solutions, or cleaning compounds. In accordance with the Hazardous Communication Standard in 40 CFR 1910.1200, personnel should be informed of safety precautions required for safe handling of such chemicals. A Material Safety Data Sheet (MSDS) for each chemical should be appended to the site-specific Health and Safety Plan described in a previous section. Training regarding the location and use of the closest eyewash and safety shower should also be provided.

Safety Procedures

- Lockout/tagout — When any energized equipment is taken out of service for maintenance or inspection, the lockout/tagout procedure will be followed to minimize the possibility of accidental energizing of equipment while maintenance is being performed. Applicable OSHA standards for lockout/tagout, presented in 29 CFR 1910.147, apply and are included in Appendix B.
- Line breakage — The line-breaking procedure described in Appendix B will be followed whenever permanent, fixed, flexible, or temporary piping is separated for maintenance or repair work. This procedure minimizes the possibility of injuries, spills, fires, and explosions during operations involving disconnection and reconnection of a line to any ECRS equipment.
- Confined space entry — The empty or partially filled ECRS soil tank meets the OSHA definition of a confined space when the gate of the tank is closed and personnel are working in it at any depth where a ladder is needed to exit the tank. If the materials being worked on or equipment being used inside the tank present a hazard to the personnel, such as the fumes generated by gasoline-fueled soil compactors, then the tank is defined as a permit confined space. Applicable OSHA standards for confined space entry are presented in 29 CFR 1910.146 and included in Appendix B.

2.7 OPERATING PROCEDURES AND MAINTENANCE

2.7.1   Routine Operating Procedures

Normal operating procedures for the ECRS Unit 1 may be implemented after installation of the equipment is completed and the unit has been inspected for safety. Setup procedures not covered in this monograph include placement of the soil tank; packing of the permeable media; connection of electrical service; installation of piping, instrumentation, and equipment; and positioning of air and water treatment units. At the discretion of the principal investigator, normal operations may also include various analytical tests to evaluate the performance of the air injection and SVE systems and the tightness of the sealed soil tank.

For the purpose of this discussion, ECRS Unit 1 has been divided into three subsystems: water, air injection, and SVE. Depending upon individual experimental requirements, the systems may be operated in several modes. In general, operation of each subsystem will involve three general steps, pre-startup inspection, startup procedure, and post-startup inspection. Shutdown procedures are also specified in order to remove the system from operation for maintenance, repair, or completion of a phase of testing.

Water Circulation System

After completing the pre-startup inspection, the water circulation system may be employed in several operating modes, as discussed below (see also Table 2.6):

- Initial filling of the soil tank — Prior to conducting technology evaluation, the soil media in the tank may be filled with water to flush out fines or increase compaction. Rapid filling can liquefy the media in the tank and disrupt compaction and artificial stratification created during soil packing. A filling rate of 0.5 gpm is suggested as a rate that is sufficiently low to ensure minimum disturbance of sand-sized grains in the media.
- Normal and reverse flow — Normal flow is defined as water recirculating from the high pressure (water injection) end of the ECRS soil tank to the low pressure (water withdrawal) end at the door, and reverse flow is flow from the door end back to the opposite wall.
- Draining the ECRS soil tank — Disposal of the soil media used for experiments will be facilitated by first draining the pore water from the ECRS tank.

Table 2.6 Groundwater System Operating Procedures — Unit 1

1.0 Pre-startup inspection

1.1 *ECRS Tank:* Visually inspect the following for loose connections, signs of wear, breakage, or other visible problems: i) ECRS tank cover, door, and connections; ii) air sparge system components; and iii) SVE system components.

1.2 *Process Equipment Skid:* Inspect process equipment skid components for signs of wear, breakage, loose wiring or hose connections, or other visible problems. Ensure that lockout/tagout service shutoff electrical switches on back of control panels are in the "on" position.

1.3 *Water Tanks and Hoses:* Inspect water tanks and associated hoses for integrity and signs of water or leakage. Check hose connections and valve positions for intended mode of operation.

1.4 *Instrumentation Building:* Inspect the general internal and external condition of instrumentation building. Check proper connection of electrical cables above transformer on back of the building.

1.5 *Electrical Service and Grounding:* Inspect electrical power cables, connections, and grounding cables for proper placement and integrity.

2.0 Filling the ECRS tank

2.1 *Water Tanks:* Ensure that water tanks are filled with water to levels corresponding to water volume necessary for filling the ECRS tank to the desired fill level.

2.2 *Open Valve Settings:* Adjust the following valves to the fully open position:

ECRS tank:	1-HV-181	1-HV-182
Interior process equipment skid:	1-HV-271	1-HV-412
	1-HV-272	1-HV-415
Exterior process equipment skid:	1-HV-232	1-HV-234
	1-HV-236	1-HV-252
	1-HV-253	1-HV-255
	1-HV-256	1-HV-258
	1-HV-259	
Water Reservoir tanks:	1-HV-432	1-HV-434

2.3 *Closed Valve Settings:* Close all valves except those required to be open as shown in Step 2.2.

2.4 *Pump Controller Setting:* Set potentiometer control on 1-P-02 pump controller box (lower of two Bronco II controllers) to 0. Set start/stop switch to "stop."

2.5 *Pump Controller Setting:* Set potentiometer control on 1-P-01 pump controller box (upper of two Bronco II controllers) to 50%. Set start/stop switch to "start." Set run/jog switch to "run."

2.6 *Pump Startup:* Push Pump 1-P-01 power switch on control box to "on" position to activate pump.

2.7 *Pump Rate Adjustment:* Refer to volume/rate gauge 1-FE-422 for actual water flow rate, and adjust the 1-P-01 pump controller potentiometer to achieve the desired pumping rate (suggested rate: 0.5 gpm).

3.0 Normal water flow

3.1 *Open Valve Settings:* Adjust the following valves to fully open position (other valves should be closed):

ECRS tank:	1-HV-181	1-HV-182
Interior process equipment skid:	1-HV-271	1-HV-411
	1-HV-272	1-HV-413
	1-HV-274	1-HV-415
	1-HV-275	
Exterior process equipment skid:	1-HV-232	1-HV-255
	1-HV-234	1-HV-256
	1-HV-252	1-HV-258
	1-HV-253	1-HV-259

continued

Table 2.6 (continued) Groundwater System Operating Procedures — Unit 1

3.2   *Closed Valve Settings:* Close all valves except those that are required to be open as shown in Step 1.

3.3.   *Pump Controller Setting:* Set potentiometer control on 1-P-01 pump controller box (upper of two Bronco II controllers) to 50%. Set start/stop switch to "start." Set run/jog switch to "run."

3.4   *Pump Controller Setting:* Set potentiometer control on 1-P-02 pump (lower of two Bronco II controllers) to 50%. Set start/stop switch to "start." Set run/jog switch to "run."

3.5   *Pump Startup:* Push power switch on the power control box to "on" position to activate pumps.

3.6   *Pump Rate Adjustment:* Referring to the flow meter (1-FIT-422) for actual water flow rate, adjust the 1-P-01 and 1-P-02 pump controller potentiometers to achieve the desired pumping rate (suggested rate: 1.0 gpm).

4.0   Reverse water flow

4.1   *Open Valve Settings:* Adjust the following valves to fully open position (other valves should be closed):

ECRS tank:	1-HV-181	1-HV-182
Interior process equipment skid:	1-HV-271	1-HV-411
	1-HV-272	1-HV-413
	1-HV-274	1-HV-415
	1-HV-275	
Exterior process equipment skid:	1-HV-231	1-HV-255
	1-HV-235	1-HV-256
	1-HV-252	1-HV-258
	1-HV-253	1-HV-259

4.2   *Closed Valve Settings:* Close all valves except those that are required to be open as shown in Step 1.

4.3   *Pump Controller Setting:* Set potentiometer control on 1-P-01 pump controller box (upper of two Bronco II controllers) to 50%. Set start/stop switch to "start." Set run/jog switch to "run."

4.4   *Pump Controller Setting:* Set potentiometer control on 1-P-02 pump (lower of two Bronco II controllers) to 50%. Set start/stop switch to "start." Set run/jog switch to "run."

4.5   *Pump Startup:* Push power switch on power control box to "on" position to activate pumps.

4.6   *Pump Rate Adjustment:* Refer to flow meter (1-FIT-422) for actual water flow rate, adjust 1-P-01 and 1-P-02 pump controller potentiometers to achieve desired pumping rate (suggested rate: 1.0 gpm).

5.0   Draining the ECRS tank

5.1   *Water Tank Status:* Check that water tanks are empty to extent corresponding to water volume necessary to drain ECRS soil tank.

5.2   *Open Valve Settings:* Adjust the following valves to fully open position (other valves should be closed).

ECRS tank:	1-HV-181	1-HV-182
Interior process equipment skid:	1-HV-271	1-HV-413
	1-HV-272	1-HV-415
	1-HV-411	
Exterior process equipment skid:	1-HV-233	1-HV-255
	1-HV-234	1-HV-256
	1-HV-252	1-HV-258
	1-HV-253	1-HV-259

5.3   *Closed Valve Settings:* Close all valves except those that are required to be open as shown in Step 2.

5.4   *Pump Controller Setting:* Set potentiometer control on 1-P-02 pump controller box (lower of two Bronco II controllers) to 0. Set start/stop switch to "stop."

5.5   *Pump Controller Setting:* Set potentiometer control on 1-P-01 pump controller box (upper of two Bronco II controllers) to 50%. Set start/stop switch to "start." Set run/jog switch to "run."

Table 2.6 (continued) Groundwater System Operating Procedures — Unit 1

5.6   *Pump Startup:* Push power switch on pump 1-P-01 control box to "on" position to activate pump.

5.7   *Pump Rate Adjustment:* Refer to flow meter 1-FIT-422 for actual water flow rate, and adjust the 1-P-01 pump controller potentiometer to achieve the desired pumping rate (suggested rate: 0.5 gpm).

6.0   Post-startup inspection

6.1   *Water Circulation System:* Check water flow meter (1-FE-422) and pressure gauges (1-DPG-421 and 1-DPG-261) to ensure pump system is working properly and appropriate flow rate is established.

6.2   *Pump Priming:* If gear pumps are dry, they may require priming. To prime pumps, partially close water pump discharge valves (1-HV-272 and 1-HV-275) for 10–20 sec until water flow is established in each pump. Fully reopen each valve after priming.

7.0   Normal shutdown

7.1   *System Shut-Down:* Switch pump power switch to "off" position to deactivate gear pumps.

7.2   *Mothballing:* If pump will be shut down for extended time, see vendor information regarding servicing.

Note: These procedures are presented as originally configured. Some revision may have occurred if equipment and systems were modified during retrofitting.

SVE System

The SVE system has only one operational mode (i.e., on or off, Table 2.7). Adjustments may be made to the applied vacuum and air flow rates through the use of pressure regulators on the process equipment skid and valves located within the manifold assemblies of the system.

Table 2.7 Soil Vapor Extraction System Operating Procedures — Unit 1

1.0   Pre-startup inspection

1.1   *ECRS Tank:* Visually inspect the following for loose connections, signs of wear, breakage, or other visible problems: i) ECRS tank cover, door, and connections; ii) air sparge system components; and iii) SVE system components.

1.2   *Process Equipment Skid:* Inspect process equipment skid components for signs of wear, breakage, loose wiring or hose connections, or other visible problems. Ensure that lockout/tagout service shutoff electrical switches on back of control panels are in the "on" position.

1.3   *Water Tanks and Hoses:* Inspect water tanks and associated hoses for integrity and signs of water or leakage. Check hose connections and valve positions for intended mode of operation.

1.4   *Instrumentation Building:* Inspect the general internal and external condition of instrumentation building. Check proper connection of electrical cables above transformer on back of the building.

1.5   *Electrical Service and Grounding:* Inspect electrical power cables, connections, and grounding cables for proper placement and integrity.

2.0   Soil vapor extraction system startup

2.1   *Sample Ports on ECRS Manifold:* Check the sample ports in the SVE manifold on the ECRS tank to ensure that port valves are closed and fittings capped.
Close the following valves: 2-HV-140   2-HV-144
                                         2-HV-142   2-HV-146

2.2   *Sample Ports on Process Equipment Skid:* Check the sample port near the intake of the SVE blower (2-HV-172) to ensure that valve is closed and fitting capped.

continued

Table 2.7 (continued) Soil Vapor Extraction System Operating Procedures — Unit 1

2.3   *Open Valve Settings:* Adjust the following valves to the fully open position:
       Exterior Process Equipment Skid: 1-HV-237
       SVE Manifold: 2-HV-141
                                               2-HV-143
                                               2-HV-145
                                               2-HV-147
       ECRS Tank: 2-HV-149

2.4   *Control Panel Settings:* Use *SITEPRO* controller for SVE startup, as follows:
       Set "SVE Power" and "Sparge Power" switches off.
       Turn "Control Power" switch to "on" to activate panel and switches.
       Set "SVE" Blower Control" switch to "run."
       Set "Sparge Pump Control" to "auto."

2.5   *Control Panel Test:* Press "System Test" to initiate *SITEPRO* panel self test. "System
       Fault" indicator will blink if a problem is found.

2.6   *Start SVE Blower:* Set "SVE Power" switch to "on." If blower does not start, check
       lockout/tagout service switch on back of *SITEPRO* panel; make sure that it is not set
       to "off." If switch is off, execute SVE System Shutdown procedure and then return to
       this procedure, beginning at Step 2.4.

3.0   Post-startup inspection

3.1   *SVE System Check:* Check pressure gauges, temperature gauges, and flow meters
       to ensure that the SVE system is working properly and that an appropriate flow rate
       is established.

4.0   Normal shutdown

4.1   *Sparge System Shutdown:* Using *SITEPRO* control panel, first set "Sparge Power"
       switch to "off."

4.2   *SVE System Shutdown:* Set "SVE Power" switch to "off" to deactivate the SVE System.

4.3   *Power Off:* For complete shut-down, set "Control Power" to "off." This action will eliminate
       all power to the panel and system switches in it.

4.4   *Closed Valve Settings:* Close the following valves:
       ECRS manifold valves: 2-HV-141   2-HV-145
                                        2-HV-143   2-HV-147

Note: These procedures are presented as originally configured. Some revision may have
       occurred if equipment and systems were modified during retrofitting.

Air Injection System

The air injection system has only one operation mode (i.e., on or off, Table 2.8). Adjustments may be made to the applied pressure and air flow rates through the use of pressure regulators on the process equipment skid and valves located within the manifold assemblies of the system. Contaminants may be introduced into the soil through a wellpoint installed in the soil. Valves 3-HV-104 and/or 3-HV-108 may be used as injection ports (Appendix A, Figure A3.1).

2.7.2   Non-Routine Operating Procedures

General Emergency

Non-routine operating procedures should be implemented in the event of sudden or unplanned changes such as electrical failure, fire, or severe weather. Characterization of these situations and appropriate actions are provided in Table 2.9, under Shutdown Procedures.

Table 2.8 Air Injection System Operating Procedures — Unit 1

1.0   Pre-startup inspection

1.1   *ECRS Tank:* Visually inspect the following for loose connections, signs of wear, breakage, or other visible problems: i) ECRS tank cover, door, and connections; ii) air sparge system components; and iii) SVE system components.

1.2   *Process Equipment Skid:* Inspect process equipment skid components for signs of wear, breakage, loose wiring or hose connections, or other visible problems. Ensure that lockout/tagout service shutoff electrical switches on back of control panels are in the "on" position.

1.3   *Water Tanks and Hoses:* Inspect water tanks and associated hoses for integrity and signs of water, leakage. Check hose connections and valve positions for intended mode of operation.

1.4   *Instrumentation Building:* Inspect the general internal and external condition of instrumentation building. Check proper connection of electrical cables above transformer on back of the building.

1.5   *Electrical Service and Grounding:* Inspect electrical power cables, connections, and grounding cables for proper placement and integrity.

2.0   Soil vapor extraction system startup

2.1   *Sample Ports on ECRS Manifold:* Check the sample ports in the SVE manifold on the ECRS tank to ensure that port valves are closed and fittings capped.
Close the following valves: 3-HV-102   3-HV-109
     3-HV-104   3-HV-154
     3-HV-106

2.2   *Open Valve Settings:* Adjust the following valves to the fully open position:
Air Sparging Manifold: 3-HV-101   3-HV-105
     3-HV-103   3-HV-107
Process Equipment Skid: 3-HV-151

2.3   *Control Panel Settings:* Use *SITEPRO* controller for Air Sparging startup by setting "Sparge Pump Control" to "auto."

2.4   *Control Panel Settings:* Use *SITEPRO* controller for SVE startup, as follows:
Set "SVE Power" and "Sparge Power" switches to "off."
Turn "Control Power" switch to "on" to activate panel and switches.
Set "SVE Blower Control" switch to "run."
Set "Sparge Pump Control" to "auto."

3.0   Post-startup inspection

3.1   *Air Sparging System Check:* Check pressure gauges, temperature gauges, and flow meters to ensure that the air sparging system is working properly and that an appropriate flow rate is established.

4.0   Normal shutdown

4.1   *Sparge System Shutdown:* Using *SITEPRO* control panel, first set "Sparge Power" switch to "off."

4.2   *SVE System Shutdown:* Set "SVE Power" switch to "off" to deactivate SVE System.

4.4   *Closed Valve Settings:* Close the following valves:
Air Sparging Manifold: 3-HV-101   3-HV-105
     3-HV-103   3-HV-107
Process Equipment Skid: 3-HV-151

Note: These procedures are presented as originally configured. Some revision may have occurred if equipment and systems were modified during retrofitting.

Table 2.9 System Shutdown Procedures — Unit 1

1.0   General emergency shutdown

    1.1   *General System Shutdown:* Deactivate ECRS components as follows:

Sparging System:	Set "Sparge Power" to "off."
SVE System:	Set "SVE Power" to "off."
Control Panel:	Set "Control Power" to "off."
Water Pumps:	Shut down pump 1-P-01.
	Shut down pump 1-P-02.
	Set "Pump Switch" to "off."

2.0   Startup after electrical failure shutdown

    2.1   *Additional Pre-Startup Inspection:* Since power failures are often associated with storm activities, it is recommended that the system be examined for damage related to severe weather.

    2.2   *System Restart:* Reactivate ECRS components as follows:
Start up the SVE system.
Start up the soil sparging system.
Start up the water circulation system.

3.0   Fire response and shutdown

    3.1   *Power Off:* Turn the power "off" at the main junction box located behind the instrument building.
Warning: *Note that if it becomes necessary to fight a fire at the tank system, it is important to keep the fire away from the ECRS tank; the vapors held underneath the cover are potentially flammable and/or explosive.*

4.0   Normal shutdown

    4.1   *Sparge System Shutdown:* Using *SITEPRO* control panel, first set "Sparge Power" switch to "off."

    4.2   *SVE System Shutdown:* Set "SVE Power" switch to "off" to deactivate the SVE System.

    4.3   *Complete Shutdown:* Set "Control Power" to "off." This will eliminate all power to panel and system switches.

    4.4   *Closed Valve Settings:* Close all valves.

Note: These procedures are presented as originally configured. Some revision may have occurred if equipment and systems were modified during retrofitting.

Freeze Protection Plan

The ECRS system is not designed to operate under freezing conditions and has no freeze protection on the tanks, piping, or instrumentation. The procedure detailed in Table 2.10 is designed to prevent or minimize freeze damage from sustained cold temperatures. Water stored in excess of 1,000 gal in the two water storage reservoirs (1-T-02, 03) is not expected to freeze during short-term cold temperatures. If temperatures are predicted to remain below 0°C for longer than 24 hr, the procedures outlined in Table 2.10 should be implemented.

2.7.3   Waste Handling and Disposal

Waste materials generated during experiments conducted in the ECRS will include the media in the soil tank, used personnel protective equipment (PPE), spent GAC, and used filters for air and water. Researchers will determine the appropriate disposal practices based upon the types

Table 2.10 Freeze Protection Procedures — Unit 1

1.0 Drain water

1.1 *Water Pump:* Switch water pump power off.

1.2 *Drain Water Lines:* Open drain valves to 1-PIT-221 and 1-PIT-223. Drain the lines into a bucket, dispose of water into one of the water holding tanks 1-T-02 or 1-T-03.

1.3 *Closed Valve Settings:* Close the following valves:

Water Holding Tank Inlet Valves:	1-HV-431	1-HV-433
Water Holding Tank Outlet Values:	1-HV-432	1-HV-434
ECRS Tank:	1-HV-181	1-HV-182
Liquid-phase GAC:	1-HV-411	1-HV-413

1.4 *Drain Water Hoses:* Disconnect the water recirculation hoses from tanks 1-T-02, 1-T-03, and 1-T-01, drain water into buckets, and dispose of water into one of the water holding tanks (1-T-02 or 1-T-03).

1.5 *Isolate Process Equipment Skid:* Close inlet valve 1-HV-234 to the water manifold on the process equipment skid.

1.6 *Drain Water Manifold:* Open drain valve 1-HV-238 and vent 1-HV-273. Catch water in buckets and dispose in one of the water storage tanks (1-T-02 or 1-T-03).

1.7 *Isolate Water Manifold:* Close the outlet valve of the water manifold 1-HV-238.

1.8 *Drain Water Filters:* Open drain valves of water filters 1-F-01, 1-F-02, and 1-F-03. Catch water in buckets, and dispose in one of the water holding tanks (1-T-02 or 1-T03).

Note: These procedures are presented as originally configured. Some revision may have occurred if equipment and systems were modified during retrofitting.

of contaminants and their concentrations. General guidelines and applicable regulations are outlined below.

Waste Media

The principal investigator is responsible for contacting a disposal facility regarding requirements for waste profiling, manifesting, transportation, disposal, and decontamination of the soil tank. The researcher is responsible for ensuring that disposal is conducted in accordance with regulations of the host facility as well as applicable federal, state, and local regulations.

Used PPE and Used Filters

If the principal investigator determines that used PPE and used filters have been contaminated with hazardous waste, the PPE and/or filters should be retained on site in a labeled DOT-certified 55-gal drum and then disposed of in accordance with the regulations of the host facility as well as applicable federal, state, and local regulations (RCRA 40 CFR 260).

Spent GAC

Spent GAC will be returned to the supplier or manufacturer for regeneration or disposal. Contact the supplier for information regarding waste profiling requirements, manifesting, transportation, and recycling facilities.

2.7.4 Troubleshooting and Maintenance

Troubleshooting procedures for the various components of the system are provided in manuals supplied by the manufacturer and included in appendices in AATDF Report TR-97-1.

Table 2.11 Required Spare Parts — Unit 1

Number	Name	Parts/Supplies	Vendor	Phone
3-B-02	Soil Sparging Blower	Flushing solvent, vanes, bearings, gaskets	Gast Manufacturing Corp.	(201) 933-8484
2.B-01	SVE Blower	Gear oil, bearing grease, shaft seals, bearings, drive belt, timing gears, impellers	Allen-Stuart Equipment Company	(713) 896-6510
2-F-04	SVE Blower Pre-Filter	Filter element	Stoddard Silencers, Inc.	(708) 223-8636
3-F-05	Soil Sparging Blower Pre-Filter	Filter element	Gast Manufacturing Corp.	(201) 933-8484
1-F-01	Water Recirculation Pump Coarse Pre-Filter	Filter housing gaskets, filter cartridges	CUNO Inc.	(203) 237-5541
1-F-02 1-F-03	Water Recirculation Pump Fine Pre-Filter	Filter housing gaskets, filter cartridges	CUNO Inc	(203) 237-5541
1-GAC-01 1-GAC-02 2-GAC-03 2-GAC-04	Granular Activated Carbon	Replacement GAC units	Calgon Corp.	(713) 486-6557
1-FIT-422 2-FIT-143 2-FIT-146 3-FIT-121 3-FIT-125	Flow Meter and Transmitter	Fuse	Endress & Hauser	(317) 535-7138
1-PIT-221 1-PIT-223 2-PIT-141 2-PIT-144 3-PIT-122 3-PIT-124	Pressure Transmitter	O-rings, miscellaneous	Endress & Hauser	(317) 535-7138
DL-101	Data Logger	D-size alkaline batteries, desiccant	Local retailer	
1-P-01 1-P-02	Groundwater Recirculation Pumps	Gear service kit	Micropump, Inc.	(360) 253-2008
NA	Sampling Pump	Tubing and pump parts	Cole-Parmer Instrument Company	(708) 647-7600
NA	Motor Speed Controller	Power module, switches, fuse	Warner Electric	(803) 286-6927
NA	Air Compressor	Miscellaneous	Thomas Industries	(414) 457-4891
NA	Roller Skids	Rollers and miscellaneous	Multiton	(804) 737-7400

In order to maintain the ECRS in optimal condition, timely maintenance and repair are required for system components. A list of suggested spare parts and sources are provided in Table 2.11. Sediment contained in the recirculated water from the soil tank may necessitate changing filters on a regular basis. Procedures for changing water filters are provided in Table 2.12.

2.8 SETUP — BACKGROUND AND SCHEDULE

Instrumentation and equipment for ECRS Unit 1 were initially assembled at the GSI warehouse in Houston, TX. The soil tank was manufactured and modified by Galbreath, Inc., Mansfield, TX. The instrumentation building was fabricated by Bebco, Texas City, TX. The process equipment

Table 2.12 Filter Change Procedure — Unit 1

1.0   Change coarse water recirculation filter (1-F-01)

1.1   *Personnel Protective Equipment (PPE):* Put on appropriate PPE.

1.2   *Isolate Filter:* Close filter isolation valves, as follows:
Downstream Isolation Valve   1-HV-253
Upstream Isolation Valve   1-HV-252

1.3   *Drain Water Filter:* Open drain valve at top of water filter, and allow all water to drain from filter housing. Catch water in a bucket.

1.4   *Change Filter:* While supporting filter housing with one hand, loosen retaining ring at top of filter with other hand. Remove filter housing, and remove and retain filter for disposal. Place new filter in housing and replace filter housing. Tighten retaining ring.

1.5   *Reconnect Filter:* Open filter isolation valves, as follows:
Upstream Isolation Valve   1-HV-252
Downstream Isolation Valve   1-HV-253

1.6   *Disposal:* Dispose of collected water in a water holding tank (1-T-02, 03). Dispose of used filters and PPE.

2.0   Change fine water recirculation filter (1-F-02)

2.1   *Personnel Protective Equipment (PPE):* Put on appropriate PPE.

2.2   *Isolate Filter:* Close filter isolation valves, as follows:
Downstream Isolation Valve   1-HV-256
Upstream Isolation Valve   1-HV-255

2.3   *Drain Water Filter:* Open drain valve at top of water filter, and allow all water to drain from filter housing. Catch water in a bucket.

2.4   *Change Filter:* While supporting filter housing with one hand, loosen retaining ring at top of filter with other hand. Remove filter housing, and remove and retain filter for disposal. Place new filter in housing and replace filter housing. Tighten retaining ring.

2.5   *Reconnect Filter:* Open filter isolation valves, as follows:
Upstream Isolation Valve   1-HV-255
Downstream Isolation Valve   1-HV-256

2.6   *Disposal:* Dispose of collected water in a water holding tank (1-T-02, 03). Dispose of used filters and PPE.

3.0   Change fine water recirculation filter (1-F-02)

3.1   *Personnel Protective Equipment (PPE):* Put on appropriate PPE.

3.2   *Isolate Filter:* Close filter isolation valves, as follows:
Downstream Isolation Valve   1-HV-258
Upstream Isolation Valve   1-HV-259

3.3   *Drain Water Filter:* Open drain valve at top of water filter, and allow all water to drain from filter housing. Catch water in a bucket.

3.4   *Change Filter:* While supporting filter housing with one hand, loosen retaining ring at top of filter with other hand. Remove filter housing. Remove and retain filter for disposal. Place new filter in housing and replace filter housing. Tighten retaining ring.

3.5   *Reconnect Filter:* Open filter isolation valves, as follows:
Upstream Isolation Valve   1-HV-259
Downstream Isolation Valve   1-HV-258

3.6   *Disposal:* Dispose of collected water in one of the water holding tanks (1-T-02, 03). Dispose of used filters and PPE.

Note: These procedures are presented as originally configured. Some revision may have occurred if equipment and systems were modified during retrofitting.

skid was fabricated at the GSI warehouse of components obtained from ORS Environmental Systems, Greenville, NH. GSI also fabricated the manifolds and harnesses for meters and gauges on the front of the tank. This assembly occurred at the warehouse from December 1, 1995 to February 26, 1996.

During this period, a Letter of Agreement was signed between Rice University and Shell Development Co. for the purposes of operating ECRS Unit 1 at Shell. The agreement outlined the roles and responsibilities of both Rice University (AATDF) and Shell during the project (Table 2.13).

Table 2.13 Responsibilities of Rice and Shell during use of ECRS Unit 1

Rice University (AATDF)	Shell Development Co.
• provide project manager for systems test	• provide project manager for air sparging test
• contract for process design drawings for Unit 1	• provide site security and access for AATDF personnel and contractors
• contract for equipment fabrication and assembly off-site and reassembly of components at Shell	• provide paved concrete area with spill protection for tank
• provide all modular components and data acquisition system for Unit 1	• provide receiving and storage facilities for Unit 1 equipment and soils prior to setup
• analyze soils for tank packing (grain size and mineralogy)	• provide utilities and appropriate hookups for equipment
• provide tank soil packing, including equipment and operator	• install stainless steel gas lines from the bottle rack to the GC in the instrumentation building
• provide first set of carbon drums for water and air treatment	• provide a health and safety inspection after setup and before operation
• operate the systems test	• operate the unit during all Shell tests
• dismantle and decontaminate the equipment after the air sparging test	• dispose of soil, rinse soil tank after test, and dispose of wastewater and used supplies
• pack the equipment and ship to off-site location	• share data on systems test and on long-term operation and maintenance of system
• provide scheduled and unscheduled system maintenance	

The equipment was disassembled at the GSI warehouse and packed onto a low-boy trailer for shipping on the morning of February 27, 1996. All of the equipment was moved using a heavy-duty extending arm forklift. The soil tank was positioned on the center of the trailer and the process equipment skid was placed inside the tank through its rear door. One water reservoir was positioned on the back of the trailer, behind the soil tank. The instrumentation building was then moved onto the trailer in front of the soil tank, and the second water reservoir was placed in front of the building, behind the truck cab. Each piece of equipment was strapped down, either through the forklift channels (soil tank, instrumentation building) or across the top (water reservoirs). The process equipment skid was secured inside the soil tank using chains, two A-frames of PVC pipe, and cardboard bumpers to prevent scratching the epoxy resin tank liner. A tarp was strapped across the soil tank. Loading and securing equipment modules on the trailer required only 3 hours.

The equipment modules on the low-boy trailer and a rented forklift on a separate trailer were delivered to Shell Westhollow Technology Center the same day. The ECRS modules were unloaded at Shell using the forklift and positioned in the EC building loading dock area. Soil had been delivered and stockpiled under tarps in the loading dock driveway earlier that day.

Soil packing was completed in 2 days, on March 1 and 4, 1996. One day would have been sufficient to pack the tank and place the sampling frits, but additional soil had to be ordered to complete the job, due to achieving tighter-than-calculated packing of the soil. Neutron soil density measurements were collected by Fugro at two depths during soil packing (see Appendix C for data).

The process equipment skid and instrumentation building were then moved into position near the soil tank using the forklift. The empty water reservoirs were light enough for one person to roll into place next to the soil tank. Technicians attached sample-tube connectors and manifolds for meters and gauges onto the soil tank. The sparge well, SVE wells, and stainless steel contaminant injection tubing were installed in the soil tank on March 5, 1996, and three soil cores were collected for analyses of air and water permeability (see Appendix C for data). The tank top was then sealed with silicon caulk, angle iron, and c-clamps. Connections were completed for electrical cables, water circuit, carbon treatment system, and heated sample line for the hydrocarbon vapor analyzer. Pressure and flow gauges were attached to the manifolds on the tank on March 7, 1996. The ECRS unit passed inspection by a Shell safety team on March 8, 1996, and was authorized to be connected to the main power. The water pumps were turned on, and the soil pack was flooded to begin flushing silts and clays from the matrix.

After these initial operations, flow controls were installed to link the outputs of the compressor and blower to maintain a consistent negative pressure in the tank. The pressure control and relief valves were installed on April 10, 1996.

During this period, the mechanical seal between the HDPE top and the tank lip proved to be incomplete and consideration was given to a variety of gaskets, caulks, and fabrics. The seal was successfully improved by mechanical smoothing of the tank lip after filling low areas with epoxy, addition of a one-piece neoprene gasket, substitution of vacuum grease for silicon caulk, and more closely spaced c-clamps on the angle iron. The tank passed a helium leak test on May 16, 1996 (Figure 6.2 in Chapter 6, and Appendix C, Figure C8.0).

2.9 SOIL PACKING AND WELL INSTALLATION

The ECRS soil tank holds 27 yd^3 of material. When the tank is packed for groundwater experiments, sand pack is placed in the water injection/recovery galleries at each end of the tank to cover the horizontal well screen. The sand pack thickness can range from 2 to 5.5 ft, depending upon the preference for a limited sand pack around the wells or for a full injection gallery. The sand grain-size is determined in relation to the main soil pack grain-size and the slot size of the well screen. The soil pack is designed by the researcher and placed in the tank in lifts, which are tamped with a power compactor. The researcher can also set well screens and sand pack in the soil matrix. A gravel pack is then placed in the top 6 in. of the tank, from the surface of the soil to the tank lip, both to support the fabric top and to act as a screen around the horizontal SVE injection/ extraction wells.

The first soil pack in ECRS Unit 1 was designed by the researcher to simulate subsurface heterogeneity and its effects on air-sparging implementation (Figures 2.1 and 2.2). Prior to placing soil in the tank, a horizontal well screen (schedule 40 PVC, slot size 0.01 in.) was fitted in the bottom of each end of the tank in the injection/recovery zones. Plywood dividers were placed in the vertical u-channel, and sand was poured around the well screen to a depth of approximately 2 ft from the floor of the tank. The sand pack size (20/40 mesh) was selected based upon the grain size of the soil matrix and the screen slot size.

To select the soils, samples were collected from a local distributor and analyzed for grain size and mineralogy by a soils laboratory. The grain size data are shown in Table 2.14, and all data from these soil analyses are included in Appendix C. The researcher selected a relatively clean medium to fine grain sand (Champion masonry sand) and a silty fine grain sand (Champion bank sand) for the soil pack. Grain-size analyses of the soils indicated that the masonry sand contained 4.3% coarse sand (1.0 to 0.5 mm), 41.8% medium sand (0.5 to 0.25 mm), 43.1% fine sand (0.25 to 0.1 mm), and 8.1% very fine sand. The bank sand contained 13.4 % medium sand (0.5 to 0.25 mm), 68.5% fine sand (0.25 to 0.1 mm), 6.2% very fine sand (1.0 to 0.5 mm), 1.7% silt (0.02 to 0.002 mm), and 4.3% clay (<0.002 mm). These data are included in Appendix C, as is the cross-section view of this soil pack and sampling array, shown in Appendix C, Figures C1.0 and C2.0. A combined soil compaction and waste factor of 15% was used to calculate the volumes for the soil order. The soil volumes included 23 yd^3 of masonry sand, 4 yd^3 of bank sand, 4 yd^3 of 3/8-in. gravel, and 36 bags (80 lb) of 20/40 well screen sand. These estimated volumes were sufficient except for the main soil type, which was 8 yd^3 short due to greater than expected compaction. The 36 bags of well screen sand covered the horizontal well screens to a depth of approximately 2 ft in the infiltration galleries at each end of the tank.

Side view

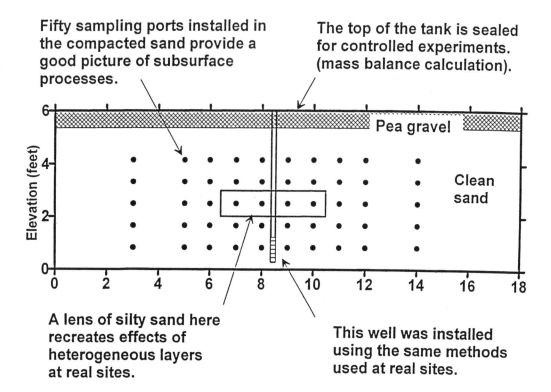

Fifty sampling ports installed in the compacted sand provide a good picture of subsurface processes.

The top of the tank is sealed for controlled experiments. (mass balance calculation).

A lens of silty sand here recreates effects of heterogeneous layers at real sites.

This well was installed using the same methods used at real sites.

Figure 2.1   ECRS soil tank setup for air sparging.

The bucket attachment on the forklift was used to move soil into the tank in 6-in. lifts, which were then compacted with a gasoline powered asphalt compactor. When the masonry sand reached a depth of approximately 2 ft, the plywood dividers were pulled from the u-channel, moved toward the center of the tank, and taped to the walls. A silty (bank) sand was packed in this approximately 1-ft thick and 4-ft wide zone to act as a potential barrier to air sparging. The barriers were then removed and soil packing continued with the masonry sand to within 6 in. of the tank lip. At the request of the researcher, ten sampling frits were laid on the soil at five selected depth horizons during this soil packing, for a total of 50 sampling frits. The frits were positioned with their screened ends along the middle line of the tank with Teflon sample tubing running to the front wall of the tank, where it was laid flat to the wall and attached to tubing sealed into two manifolds on the front of the tank.

To estimate the packed soil density neutron density measurements were made after the soil was packed. Dry density and moisture readings were taken at a depth of 6 in. below the soil surface. The dry density ranged from 89.5 to 101.2 lbs per cubic foot (pcf) (see Fugro Geosciences, Inc., data tables in Appendix C). Three soil core samples were also collected from the augered hole for the central air sparging well. The core samples were sent to Fugro Environmental, Inc. (now ENSR) for measurement of air and water permeability. Measurements were recorded for dry density, water content, water permeability, and air permeability. Dry density increased slightly toward the top of the soil pack. It ranged from 94.9 pcf at depths of 53 to 59 in. to 98.6 pcf at depths of 36 to 42 in., and 99.6 pcf at 12 to 18 in. below the soil surface. The middle core (36 to 42 in.) was taken from the reduced permeability layer in the center of the soil pack, and water and air permeability data from the cores indicated that the layer was tighter than the surrounding soil pack. Water permeability ranged from 9.3×10^{-3} cm/sec, at a depth of 53 to

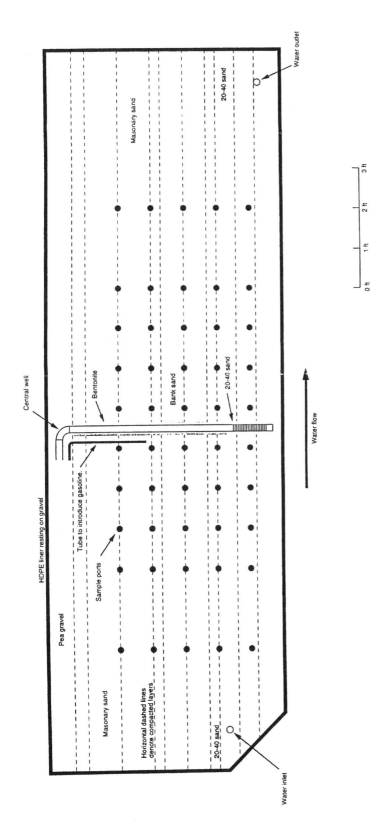

Figure 2.2 ECRS soil tank internal arrangement for air sparging.

Table 2.14 Grain-Size Distribution in Candidate Soils for Unit 1 Soil Pack at Shell

Champion Co. Soils	Sand (mm)						Silt (mm)		Clay (mm)		TEXTURE CLASS
	VC[a] 2.0–1.0 (mm)	C 1.0–0.5	M 0.5–0.25	F 0.25–0.1	VF 0.1–0.05	TOTAL 2.0–0.05	FINE 0.02–0.002	TOTAL 0.05–0.002	FINE <0.0002	TOTAL <0.002	
Champion Loam	0.3	3.8	28.7	37.1	11.4	81.3	7.8	14.4	1.9	4.3	LS
Champion Top Soil w/clay	0.6	0.8	6.0	29.4	30.1	66.9	9.9	23.5	5.4	9.6	VFSL
Champion Red Clay	0.4	7.1	41.8	27.6	2.4	79.3	2.2	3.7	13.2	17.0	SL
Champion Masonary Sand[b]	0.5	4.3	41.8	43.1	8.1	97.8	0.3	1.2	0.5	1.0	S
Champion Bank Sand[b]	0.1	0.2	13.4	68.5	6.2	88.4	1.7	3.9	4.3	8.5	LFS

[a] VC, very coarse; C, coarse; M, medium; F, fine; VF, very fine.

[b] These two soils were selected for the soil pack in ECRS Unit 1 at Shell Development Co., Houston, TX.

The masonary sand was chosen for the main portion of the soil pack. The bank sand was used for the reduced permeability layer in the center of the soil pack.

59 in., to 1.3×10^{-3} cm/sec, at a depth of 36 to 42 in. in the reduced permeability layer, and 5.9 $\times 10^{-3}$ cm/sec at 12 to 18 in. below the soil surface. Air permeability ranged from 40.7 Darcy at a depth of 53 to 59 in. to 5.0 Darcy at a depth of 36 to 42 in. in the reduced permeability layer, and 19.2 Darcy at 12 to 18 in. below the soil surface. These data are included in Appendix C.

After the neutron density testing and coring, two horizontal SVE wells were installed on the surface of the packed soil at the ends of the tank, and 6 in. of gravel was placed over the soil and wells. The fabric top was then laid over the gravel and sealed to the tank lip.

Summary of ECRS Unit 1 Operations

Unit 1 was set up at Shell Westhollow Technology Center, Houston, TX, in the spring of 1996. It underwent a safety inspection and an initial 3-month systems test of the equipment followed by a 12-month air sparging research project. Mr. Jonathan Miller, Shell, was the principal investigator on the project. His soil pack and sensor/sampler array were designed to test the effect of minor soil heterogeneities on air sparging. He also varied air sparging operating parameters and tested the effectiveness of oxygen releasing materials (ORM). Unit 1 was disassembled in August 1997. The unit was retrofitted with minor, new components and a modified tank lip for top attachment and shipped to the USAE Waterways Experiment Station (WES) in Vicksburg, MS, in January 1998, to test pilot-scale remediation technologies.

2.10 TANK COVER AND SHELTER INSTALLATION

Both soft and hard tops were considered; the soft top was selected because of its light weight. It is easy to move by hand instead of with a crane. Manways can also be added for access to the soil pack, and the soft top reduces the potential explosive hazard associated with the enclosed tank.

The original top for Unit 1 was composed of 60-mm thick Gundle high density polyethylene (HDPE) landfill liner without manways. The flat edge of the HDPE was mechanically sealed against the 5-in. wide lip of the tank with a 0.5-in. neoprene gasket and silicon caulk, and secured with angle iron and c-clamps. To improve the mechanical seal, the tank lip was prepared for placement of the gasket and caulk by smoothing the metal surface with a grinder and filling irregularities with epoxy. Vacuum grease was substituted for silicon caulk. The surface of the HDPE was covered with a tarp to reduce material degradation from incident sunlight.

To provide long-term durability, a second soft top was fabricated by Mesa Rubber Co., Houston, TX, from two-ply urethane (Mesa 6036.35 FCA, one-ply ester-based and one-ply ether-based) with low sorption and low permeability to fuel contaminants. Two carbon steel manways with Teflon bolts were designed to provide access to the soil and instrumentation after the top was sealed.

Weather Shelter for Tank

A fabric sun/rain shelter is included with the ECRS unit to shield the soil tank and water reservoirs from direct summer sunlight, heat, and rain. The shelter frame and base from Hansen Weather Port, Gunnison, CO, are constructed of galvanized steel tubing. The framework is a heavy-duty gable style with webbed arches and measures 18 ft wide, 20 ft long, and 10 ft high at the arch. The main arches are on 10-ft centers with lateral stability cables. The cover, which extends only across the roof and 1 ft down each long side, is composed of 18-oz polyester reinforced vinyl with a cable tensioning system. The base is anchored with concrete expansion bolts. The shelter packs into reusable shipping crates.

CHAPTER 3

ECRS Engineering Design and Operation — Unit 2

3.1 DESIGN PARAMETERS

The Unit 1 modules performed to design specifications based upon performance at Shell Westhollow and later quantitative system testing by GSI. AATDF solicited comments on the design and performance from Shell researchers, AATDF advisors, the GSI fabrication team headed by Thomas Reeves, and other ECRS advisors, including Paul Johnson of Arizona State University. Suggested design modifications that were implemented in the Unit 2 design included the following:

- addition of a sight glass column to view the water level in the tank
- substitution of a centrifugal pump for the rotary gear pumps to facilitate moving the filters downstream of the pump
- relocating the water filters downstream of the water pump
- substitution of a screw-type air compressor for the centrifugal air pump on Unit 1
- removing the SitePro-SpargePro Controller System
- locating all gauges and controls on one panel on the process equipment skid and removing gauges and controls from the soil tank
- adding humidity control to the instrumentation building to reduce condensation
- removing the heated sample line and thermal vapor analyzer instrumentation and substituting a portable GC
- replacing the HDPE tank top with a two-ply urethane fabric top
- using one 1,500-gal opaque water reservoir instead of two 1,200-gal reservoirs
- adding an air accumulator, supply-air tank to the process equipment skid
- addition of a portable 50-gal chemical mixing tank

AATDF contracted with GSI to assist in preparation of the detailed design and bid package for ECRS Unit 2. GSI proposed a scope of work that included the following tasks:

Task 1: Unit 2 Detailed Engineering
    Project startup
    System design: Review and modification of the Unit 1 design
    Development of design package of engineering drawings and equipment specifications
        Process flow diagram
        Engineering flow diagram
        Electrical one-line diagram
        Equipment specifications
        Fabrication drawings

Task 2: Unit 2 System Fabrication Oversight
  Bid package preparation
  Assistance to AATDF in subcontractor selection
  Fabrication oversight
Task 3: System Delivery Oversight
  Preliminary site visit
  Delivery oversight
Task 4: System Installation Oversight
Task 5: Project Documentation
  Operations and maintenance manual preparation
  Operator training

Overview of the Engineering Bid Package

The bid package prepared for fabrication of ECRS Unit 2 included instructions to bidders on the general requirements, schedule, design basis, equipment list, bid quotation form, system configuration, and the ECRS Unit 1 example. Specifications were provided for the process equipment skid, soil tank, piping, and ancillary equipment. The bid package and specifications are included in Appendix D.

Based upon the four bids received, AATDF selected Protec, Dickinson, TX, to fabricate the process equipment skid, modify a 27-yd³ rectangular steel sludge container, and purchase other materials and equipment. Separate bids were received for fabrication of the instrumentation building. The completed building was delivered to Protec for inspection during fabrication and staging of Unit 2.

The assembled ECRS Unit 2 consists of the standard four modules including an instrumentation building, process equipment skid, soil tank, and reservoirs. The insulated instrumentation building had climate control, bench space for analytical and computer equipment, an exterior bottled-gas rack, stainless steel piping for GC gases, electrical outlets, and storage space. The process equipment skid, fabricated by Protec, was equipped with a compressor, a blower, groundwater pumps, water and air filters, piping, an electrical panel, a gauge panel, and supply air tank, a portable chemical mixing tank, and system controls. The soil tank was a 27-yd³ rectangular steel sludge container purchased from Galbreath, Inc. and modified by Protec, Inc. with a system of top bolts and angle iron, pass-thru pipes with threads and caps, flanges, tubing nozzles, valves, tee-strainers, sight gauges, an upgraded door seal, and interior epoxy coating.

Unit 2 Setup and Operations Schedule

Unit 2 was set up at Arizona State University, Tempe, AZ in the spring of 1997. Dr. Paul Johnson, the principal investigator, conducted 3-D pilot-scale research on air sparging. He is a member of a large research team addressing various aspects of physical model studies of *in situ* air sparging performance. Funding for the project was provided by the USAF, SERDP, API, and AATDF/Rice University. Dr. Johnson's experiment was to test air sparging treatment methods for source zones and dissolved plumes and to perfect diagnostic tools for air sparging performance. Unit 2 was returned to Houston, TX in August 1998.

3.2 UNIT 2 ENGINEERING DIAGRAMS

The engineering design for ECRS Unit 2 was provided by Groundwater Services Inc. (GSI), Houston, TX and was critiqued and modified by the AATDF staff and advisors. Ten design figures were prepared as part of the three-volume *Operation and Maintenance Manual for ECRS Unit 2* (AATDF 1997b). Those figures are listed below and included in Appendix E.

3.3 DESCRIPTION OF SYSTEM COMPONENTS

Unit 2 of the ECRS was designed and fabricated to facilitate SVE, air injection or sparging, and water circulation subsystems. The performance of these systems is summarized in Table 3.1. A process flow diagram for the unit is shown in Appendix E, Figure E5.0. Detailed engineering flow diagrams were presented in the previous chapter and are included in Appendix D. Summary information regarding system configuration and controls is provided below.

3.3.1 Process Equipment Skid

The process equipment skid shown in Photo 3.1 contains pumps, piping, and controls to operate pilot-scale experiments in the ECRS soil tank. The front and rear isometric diagrams for the process equipment skid are illustrated in Figures 3.1 and 3.2, respectively. A list of major equipment is provided in Table 3.2, which includes a list of the two-part equipment numbers used on the design drawings (e.g., C-1, air compressor). Details regarding instrumentation and equipment installed on the process equipment skid are provided in Appendix D. In general, equipment and instrumentation on the process equipment skid can be divided into an air system, water system, and controls, as follows:

- Air System — A screw-type air compressor (C-1) and an air accumulator (V-1) provide pressurized air for injection or sparging and SVE for media in the ECRS soil tank. Particulates present in ambient air are removed by a filter that is built into the compressor unit. Prior to entering the air receiver (V-1) compressed air is treated by a coalescing filter (F-1). A carbon filter (F-4) installed downstream of the air receiver removes any remaining particulates and organic compounds. Regulated pressurized air for the accumulator tank (V-1) is used for air injection or sparging and as a motive force to a venturi eductor for SVE and to provide a utility air station on the process equipment skid. Entrained moisture and particulate matter are removed from the extracted soil vapor by means of a moisture separator vessel (V-2). The resulting air is then routed to a treatment system provided by the principal investigator.
- Water System — Water is circulated through the soil in the ECRS soil tank (T-1) using a centrifugal pump (P-1). After passing through the pump, water is filtered through spiral-wound filter elements (F-2 and F-3) to remove soil particles. Connections are provided for routing water from the soil tank to an additional treatment system for removal of organic and inorganic constituents that are provided by the principal investigator. An additional utility pump (P-2) has been mounted on the skid.
- Controls and Electrical — Two control panels (CP-1 and CP-2) enable setting system parameters, monitoring system performance, and shutdown in the event of an alarm condition (i.e., high pressure, high liquid level, or electrical motor overload). The system control schematic is depicted on the electrical ladder diagram in Appendix E, Figure E10.0. A permanent record of process variables is achieved using the data recorder (UIR-1) installed in the control panel (CP-1). Further discussion of system controls and electrical requirements is provided in Section 3.4.

Table 3.1 System Performance Data — Unit 2

System	Description	Parameter	Range
Water Recirculation	Pressure regulation of the water supply and return lines connected to the ECRS tank is employed to achieve a set gradient over the length of the tank. Water in excess of that required to maintain the defined hydraulic gradient is returned to a surge tank prior to filtration, any additional treatment (i.e., carbon adsorption to remove organic compounds), and recirculation to the upgradient end of the tank.	• Groundwater flow rate • Groundwater elevation (above bottom of tank) • Hydraulic gradient	0–8 gpm 1–4 ft 0–2 ft/ft
Air Sparging	Filtered compressed air is regulated and injected into the ECRS tank. The injection pressure is referenced to the SVE vacuum and is designed to maintain a constant differential pressure across the tank.	• Air flow rate • Injection pressure (relative to applied SVE vacuum)	0–30 scfm 0–5 psi difference
Soil Vapor Extraction (SVE)	A venturi vacuum pump driven by compressed air provides SVE capability for the system. The motive air pressure is regulated to apply a constant vacuum at the top of the ECRS tank. Entrained solids and water are removed in a centrifugal separator upstream of the vacuum pump.	• Air flow rate • SVE vacuum	0–30 scfm 0–15 in. Hg vacuum
Chemical Injection	A mixing tank and metering pump are provided to facilitate the controlled addition of water soluble tracers and chemicals to the groundwater supply stream or to specific point(s) in the process.	• Chemical injection flow rate	0–7 gph

Note: Performance data shown above was determined during a series of system tests conducted by the fabricator and the Rice representative during January 1997.

Treatment of the water circulating in the ECRS tank to remove organics is the responsibility of the researcher.

Photo 3.1   Process equipment skid — Unit 2.

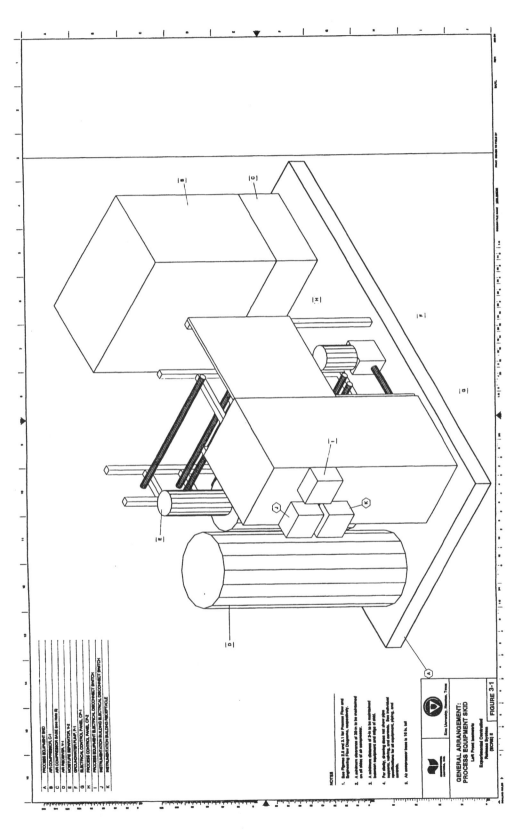

Figure 3.1 Front isometric of process equipment skid.

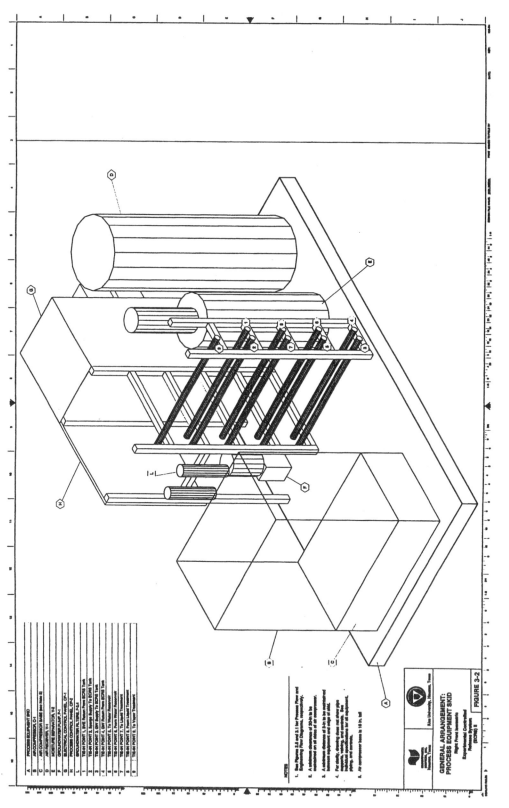

Figure 3.2 Rear isometric of process equipment skid.

Table 3.2 List of Major Equipment — Unit 2

Subassembly	Item No.	Description
Process Equipment Skid	C-1	Air Compressor
	V-1	Air Accumulator
	V-2	Moisture Separator
	F-1	Coalescing Filter
	F-2	Carbon Filter
	F-3, 4	Groundwater Filters
	P-1	Groundwater Recirculation Pump
	SP-1	SVE Venturi Pump
ECRS Tank	T-1	ECRS Tank
Ancillary Equipment	P-2	Utility Pump
	P-3	Chemical Metering Pump
	T-2	Water Reservoir
	T-3	Chemical Mixing Tank

Note: Locations of above referenced equipment are shown in Engineering Flow Diagrams for process equipment skid, ECRS tank, and ancillary equipment (Figures E.6, E.7, and E.8, respectively.)

3.3.2 ECRS Soil Tank

The soil tank (T-1) was fabricated by modifying a 27-yd^3 rectangular steel sludge container (Photo 3.2). An engineering flow diagram for the tank is provided in Appendix E, Figure E7.0. The tank has been fitted with flanges and tubing nozzles and equipped with valves, tee-strainers, and sight gauges. During installation, the tank is fitted with a flexible top to isolate the tank contents from the environment. To facilitate obtaining an accurate mass balance during experiments, the top is bolted and sealed to the tank. Nozzles through the tank walls provide for connecting soil pack instrumentation with piping and hoses to the process equipment skid. To prevent a buildup of excess air pressure or fluid level under the fabric top, a pressure relief valve that vents to the atmosphere and two high liquid-level switches have been installed.

Photo 3.2 Soil tank — Unit 2.

3.3.3 Ancillary Equipment

Additional equipment required to complete the system is shown in the engineering flow diagram in Appendix E, Figure E8.0. It includes the following: a water storage reservoir, a chemical mixing tank, and air and water treatment units. A 1,500-gal polyethylene tank (T-2) has been equipped with flexible hoses for connection to the process equipment skid. A portable 50-gal mixing tank (T-3) with attached mixer has been provided for preparing chemical amendments and tracer solutions.

3.3.4 Instrumentation Building

An insulated instrumentation building has been included with the ECRS to provide an area for operating analytical instruments, computers, and other equipment associated with research performed in the ECRS unit (Photo 3.3). The building serves as a locked storage area when researchers are not on site. The building dimensions of 8 × 8 × 8 ft are within the height and width for standard trucking without a permit. The structural steel base of the building is equipped with forklift slots and lifting lugs for a crane or for anchor straps. Access to the building is provided by a single door with panic hardware, outside key lock, and a window.

Photo 3.3   Instrumentation building — Unit 2.

The building has electrical power connections to interior and exterior lighting and electrical outlets for instrumentation. A thermostatically controlled HVAC unit and dehumidifier provide cooling and moisture control. Cable trays are installed around the interior perimeter near the ceiling for routing electrical and signal cables. Stainless steel workbenches and cabinets provide work surfaces and storage space. Unistrut installed on the exterior of the building accommodates chained storage of compressed gas cylinders.

3.3.5 Process Control and Data Acquisition

The process equipment skid is equipped with instrumentation and electronics for system control, recording and monitoring operations data, and shutdown in the event of an alarm condition. Panels house electronics and controls for the data logger; gauges, meters, and sensors; and pressure valves and gauges. These and the electrical control panel are described below:

- Data Recorder — System control and data recording are provided by a single intelligent data recorder (UIR-1) having a color strip chart recorder, a memory card capable of archiving system data in ASCII format, discrete and analog input cards, and output relay cards. Data logging software is customizable for real-time data acquisition parameters of interest for each experiment. The recorder installed in ECRS Unit 2 is equipped with math functions; counters, timers, and totalizers; derived variable calculations; and a memory card for data archiving. The recorder is capable of calculating derived variables (e.g., mass flow from volume flow, pressure, and temperature) and processing alarms (e.g., absolute high/low, deviation, increasing or decreasing rate-of-change, or digital status). An electrical ladder diagram illustrating the relationship of input/output cards to system instrumentation and alarms is provided in Appendix E, Figure E10.0. Detailed specifications and manufacturer's information are provided in AATDF Report TR-97-5.
- Electrical Control Panel — A freestanding NEMA 4 steel electrical enclosure (CP-1) mounted on the process equipment skid houses rack-mounted electrical hardware, power and constant voltage transformers, a power supply for loop-powered instrumentation, and motor starters. Enclosure cooling is provided to ensure proper operation of electrical components. Two 480 volt, 3-pole switches installed on the exterior of the electrical enclosure provide for disconnecting the process equipment skid and the instrumentation building. Fused terminal blocks have been installed for power distribution and instrumentation tie-ins. In order to provide convenient access to electrical power for portable power tools, two duplex GFCI utility outlets in weatherproof enclosures are mounted on the process equipment skid. A NEMA 4-rated instrument access door has been installed in the door of the enclosure to allow access to the data recorder (UIR-1) without opening the main enclosure door.
- Gauge Panel — A second control panel (CP-2) has been installed on the process equipment skid unit to provide convenient mounting of pressure regulator adjustment valves and pressure gauges. The panel (CP-2) has been constructed of sheet steel and hinged to provide convenient access to connections and piping on the rear of the panel.
- Instrumentation — Instruments such as sensors, flow meters, and temperature sensors provide a direct reading of conditions within the soil tank and associated piping. A list of system instrumentation is provided in Table 3.3, which also includes the instrumentation numbers used on the design diagrams. Depending on the site-specific configuration, system conditions may be monitored continuously measuring temperature, pressure, and flow rate. Additional information on system operations is provided by pressure gauges, temperature sensors, and flow meters on air and water lines to the soil tank.

Table 3.3 List of Instrumentation — Unit 2

Process Equipment Skid	Item No.	Location
Data Recorder	UIR -1	Electrical Enclosure
Level Controller	LC-1	Moisture Separator (V-2)
Differential Pressure Transmitter	LT-1	Moisture Separator (V-2)
Flow Switch	FSLL-1	Filtered Groundwater
Flow Transmitter	FIT-1	Groundwater Return from ECRS Tank
	FIT-2	Sparge Mass Flow
	FIT-3	SVE Mass Flow
Hand Switch	HS-1	Electrical Enclosure
	HS-2	Electrical Enclosure
	HS-3	Electrical Enclosure
	HS-4	Electrical Enclosure
	HS-5	Electrical Enclosure
	HS-6	Electrical Enclosure
Level Switch	LSHH-1	Moisture Separator (V-2)
	LSHH-4	Air Accumulator (V-1)
Push Button	PB-1	Electrical Enclosure

continued

Table 3.3 (continued) List of Instrumentation — Unit 2

Process Equipment Skid	Item No.	Location
Pressure Gauge	PG-1	Groundwater Gradient Control Inlet
	PG-2	Groundwater Gradient Control Outlet
	PG-3	Groundwater Pump Discharge
	PG-4	Groundwater Filter Discharge
	PG-5	Groundwater Treatment Supply
	PG-6	Sparge Supply
	PG-7	SVE Supply
	PG-8	Main Air Supply
	PG-9	Utility Air Supply
	PG-10	Groundwater Treatment Discharge
	PG-11	Utility Pump
Pressure Regulators	BPV-1	Groundwater Return Head
	PRV-1	Reference Groundwater Supply Head
	PRV-2	Reference Groundwater Return Head
	PRV-3	Air Sparge
	PRV-4	Reference SVE Vacuum
	PRV-5	Utility Air
	PRV-6	Applied Vacuum
	PRV-7	Reference Air Sparge
	PRV-8	Groundwater Supply Head
	PRV-9	Groundwater Treatment Supply
Pressure Switch	PSHH-4	Filtered Groundwater
Pressure Transmitter	PIT-1	Sparge Mass Flow
	PIT-2	SVE Mass Flow
	PIT-3	SVE Sensor
	PIT-4	Sparge Supply
	PIT-5	Groundwater Supply Pressure Sensor
	PIT-6	Groundwater Return Pressure Sensor
Sight Glass	SG-1	SVE Flow
Solenoid Valves	XV-1	Compressed Air Supply
	XV-2	Sparge Supply
	XV-3	Groundwater Reservoir Return
	XV-4	Groundwater Return from ECRS Tank
	XV-5	Groundwater Reservoir Supply
	XV-6	Groundwater Supply to ECRS Tank
Temperature Sensors	TE-1	Sparge Air Temperature
	TE-2	SVE Air Temperature
	TE-3	Groundwater Supply Temperature
	TE-4	Groundwater Return Temperature
Level Switch	LSHH-2	ECRS Tank (T-1)
	LSHH-3	ECRS Tank (T-1)
Pressure Relief Valve	PSV-1	ECRS Tank (T-1)

Note: Locations of above referenced equipment are shown in Engineering Flow Diagrams for process equipment skid and ECRS tank (Appendix E: Figures A.6 and A.7, respectively).

Specifications and manufacturers' data are provided in Appendices A and C, respectively.

3.4 SITE HAZARDS

Operation and maintenance of ECRS Unit 2 should be conducted in accordance with the site-specific project health and safety program developed by the researcher prior to startup. Suggested means for dealing with site hazards are presented here but are not comprehensive. Applicable requirements should be reviewed by the principal investigator and other relevant parties prior to implementing the health and safety program for each site. All operations and maintenance personnel working with an ECRS unit should read and be familiar with this plan prior to working with the unit. On-site responders to fire and spill conditions should be apprised of site hazards and chemicals used on site before an emergency condition arises. Information regarding site hazards posed by both ECRS units are presented in Section 2.6 and Appendix B.

Due to equipment modifications between the units, only moderately elevated noise levels on the Unit 2 process equipment skid are associated with operation of the air compressor (C-1) and the SVE venturi pump (SP-1).

3.5 SITE-SPECIFIC REQUIREMENTS

The ECRS units were designed and constructed to provide maximum flexibility in implementation of various experimental configurations. To support ECRS Unit 2, certain site requirements need to be considered, which include some minor variations from Unit 1. Those requirements are listed below.

- Workspace — A minimum area of 25 × 50 ft is required to allow adequate clearance between the components. A means of securing the site is also necessary for personnel safety and to prevent equipment theft or damage. Depending upon the chemicals and their concentrations in the ECRS tests, secondary containment or access to a chemical process sewer may be required.
- Utilities — The preferred electrical connection for the ECRS unit power supply is a 4-wire, 480 volt, 60 amp service with 3-phase conductors and a grounding conductor. Alternatively, a 208 volt, 100 amp service could be used, provided a 3-phase, 5-wire system were available with 3-phase conductors, a grounded neutral, and a grounding conductor.
- Health and Safety — A commitment is needed on the part of the principal investigator to provide controls for the health and safety of all personnel involved in the installation and operation of the ECRS. This will include familiarity with applicable workplace safety regulations, as well as personnel training and site inspections.
- Commissioning — Prior to installation of the ECRS modules on site, the researcher needs to prepare for delivery by ensuring the site has been prepared, as noted above, and that appropriate safety training has been provided for the relevant staff. During installation, there will also be the need for a laydown area for equipment prior to setup, a staging area of soil, and an access route for heavy equipment and ECRS modules.
- Decommissioning — The researcher is responsible for emptying all sensors, samplers, and soil from inside the soil tank and decontaminating the tank. Soil and rinsate should be managed in an appropriate manner. Temporary access will be needed to a convenient equipment laydown area and to a route for heavy equipment and ECRS module removal. Disposal of all wastes generated during decontamination and packing of equipment is the responsibility of the researcher.

3.6 OPERATING PROCEDURES AND MAINTENANCE

3.6.1 Routine Operating Procedures

Normal operating procedures for ECRS Unit 2 may be implemented after installation of the equipment is completed and the unit has been inspected for safety. Setup procedures not included in the operations and maintenance manual are placement of the soil tank; packing of the permeable media; connection of electrical service; installation of piping, instrumentation, and equipment; and positioning of air and water treatment units. At the discretion of the principal investigator, normal operations may also include various analytical tests to evaluate the performance of the air injection and SVE systems and the tightness of the sealed soil tank.

Table 3.4 Pre-Startup Inspection Procedures — Unit 2

1.0 Electrical Service and Grounding

1.1 *Main Power Disconnect:* Lockout main system power at service disconnect until Section 1.0 of this table is complete. Check that minimum service requirement of 60 amps at 3Ø/480V/60Hz is available.

1.2 *Process Skid Disconnect:* Lockout Process Equipment Skid power at disconnect switch on the side of the electrical control panel (CP-1).

1.3 *Instrumentation Building Disconnect:* Lockout instrumentation building power at disconnect switch on the side of the electrical control panel (CP-1).

1.4 *General Appearance:* Visually inspect the general condition of the electrical service. Check power cables, connections, and grounding cables for proper placement and integrity.

1.5 *Grounding:* Check for continuity between ground bus in electrical control panel (CP-1) and grounding location.

1.6 *Main Power:* Remove main system power lockout. Turn main system power on.

2.0 Process Equipment Skid

2.1 *General Appearance:* Visually inspect the general condition of the process equipment skid.

2.2 *Control Panels:* Inspect general condition of control panels (CP-1, CP-2). Adjust hand-loading regulators to fully unloaded positions (full counter-clockwise). Adjust panel operators to "off."

2.3 *Process Piping and Connections:* Inspect general condition of process piping. Check process connections/hoses between process equipment skid and all other equipment.

2.4 *Process Filters:* Inspect condition of process filters (F-1, F-2, F-3, F-4). Replace filter elements as necessary.

2.5 *Electrical Connections:* Inspect general condition of electrical connections. Check main power and instrumentation building power cables to see they are securely engaged in their receptacles.

2.6 *Process Equipment Power:* Remove lockout from process equipment power disconnect. Turn on process equipment power.

2.7 *Data Recorder:* Check to see that data recorder completes initialization on power-up.

2.8 *Air Compressor:* Check to see that air compressor completes diagnostic check on power-up.

3.0 Ancillary Equipment

3.1 *Groundwater Reservoir:* Visually inspect general condition of groundwater reservoir. Open block valve on the sight gauge.

3.2 *Groundwater Treatment System:* Inspect general condition of groundwater treatment system in accordance with manufacturer's recommendations.

3.3 *Vapor Treatment System:* Inspect general condition of vapor treatment system in accordance with manufacturer's recommendations.

3.4 *Chemical Injection System:* Inspect general condition of chemical injection system including mixing tank (T-3) and chemical metering pump (P-3).

4.0 ECRS Tank

4.1 *General Appearance:* Visually inspect the general condition of the ECRS tank.

4.2 *Pressure Relief Valve:* Inspect general condition of pressure relief valve. Ensure that valve exhaust is unobstructed.

4.3 *High Level Switches:* Inspect general condition of high level switches. Inspect electrical cables between high level switches and process skid.

4.4 *Flexible Top:* Inspect general condition of flexible top including integrity of seal at edge of soil tank and all nozzles.

4.5 *Process Connections:* Inspect general condition of process connections (groundwater, SVE, and injection).

4.6 *Sample Tubing and Instrumentation:* Inspect general condition of sample tubing and instrumentation.

Table 3.4 (continued) Pre-Startup Inspection Procedures — Unit 2

5.0   *Instrumentation Building*

5.1   *General Appearance:* Visually inspect internal and external condition of instrumentation building.

5.2   *Bottle Rack:* Check that all compressed gas cylinders are secure. Inspect instrument tubing and regulators for signs of damage.

5.3   *Power Distribution:* Turn off all breakers in power distribution panel including main breaker. Unplug all devices from outlets. Turn all switches to "off" position (HVAC, lighting, vent fan, etc.).

5.4   *Instrumentation Building Power:* Remove lockout from instrumentation building disconnect switch on process equipment skid. Turn on building power and main breaker in power distribution panel.

5.5   *Electrical Circuits:* Turn on one circuit at a time at the power distribution panel.

5.6   *Vent Fan:* Check proper operation of vent fan by momentarily operating switch located on exterior of building. Check vent fan filter. Replace filter if dirty.

5.7   *Climate Control:* Set thermostat for heat or cool and select a comfortable temperature. Verify that humidistat is set at 50% RH (unless another setting is desired). Check HVAC filter and replace if dirty.

6.0   *Fabric Shelter*

6.1   *General Appearance:* Visually inspect the general condition of fabric shelter.

6.2   *Installation:* Check to see that structure is appropriately anchored and secured against wind loads.

Note: These procedures are presented as originally configured. Some revision may have occurred if equipment and systems were modified during retrofitting.

For the purpose of this discussion, ECRS Unit 2 has been divided into three subsystems: water, air injection, and SVE. A pre-startup inspection procedure for each of the systems is outlined in Table 3.4. Operation of the water circulation system, the air injection or sparging system, and the SVE system are outlined in Tables 3.5 to 3.7, respectively. Normal system shutdown is described in Table 3.8.

Table 3.5 Groundwater System Operating Procedures — Unit 2

1.0   *Establishing Initial Water Level in ECRS Tank*

1.1   *Water Reservoir:* Ensure that the water reservoir is filled with water (approximately 1500 gal).

1.2   *Process Connections:* Open ECRS soil tank groundwater process connections: water inlet valve BV-33 and water return value BV-36. Open Process Equipment Skid process connections: water supply valve BV-27 and water return valve BV-8.

1.3   *Water Recirculation Pump:* Switch panel operator for water recirculation pump (HS-2) to "auto."

1.4   *Water Supply/Return:* Switch panel operator for water supply/return solenoids (HS-3) to "auto."

1.5   *Water Level:* Adjust water level in ECRS soil tank.

2.0   *Adjusting Water Level in ECRS Tank*

2.1   *Adjust Inlet Water Level:* Increase setpoint pressure at groundwater supply regulator (PRV-1) located on gauge panel (CP-2) by turning regulator adjustment knob *slowly* clockwise. Decrease setpoint pressure at groundwater supply regulator (PRV-1) located on gauge panel (CP-2) by turning regulator adjustment knob *slowly* counterclockwise.

continued

Table 3.5 (continued) Groundwater System Operating Procedures — Unit 2

2.2   *Adjust Outlet Water Level:* Increase setpoint pressure at groundwater return regulator (PRV-2) located on gauge panel (CP-2) by turning regulator adjustment knob *slowly* clockwise. Decrease setpoint pressure at groundwater return regulator (PRV-2) located on gauge panel (CP-2) by turning regulator adjustment knob *slowly* counter-clockwise.

3.0   Gradient Control

3.1   *Increase Gradient:* To increase groundwater gradient, adjust inlet or outlet water level, or both.

3.2   *Decrease Gradient:* To decrease groundwater gradient, adjust inlet or outlet water level, or both.

4.0   Draining the ECRS Tank

4.1   *Set Minimum Water Level:* Set water in soil tank to minimum by decreasing inlet and outlet water levels to minimum.

4.2   *Observe Water Level:* Observe actual water level in ECRS tank until minimum level is reached.

4.3   *Process Connections:* Close ECRS soil tank groundwater process connections: water inlet valve BV-33 and water return valve BV-36. Close process equipment skid process connections: groundwater supply valve BV-27 and groundwater return valve BV-8.

4.4   *Groundwater Supply/Return:* Switch panel operator for water supply/return solenoids (HS-3) to "off."

4.5   *Drain Residual Water:* Drain residual water from ECRS tank and hoses at low-point drains (also sample ports) at ECRS tank (BV-32, BV-37) into buckets for subsequent disposal.

5.0   Post-Startup Inspection

5.1   *Process Piping:* Inspect process piping and hoses for leaks and other indications of damage.

5.2   *Water Circulation:* Check actual water levels in soil tank and water flow rate to ensure system performance is within expected range.

6.0   Normal Shutdown

6.1   *Groundwater Supply/Return:* Switch panel operator for water supply/return solenoids (HS-3) to "off."

6.2   *Groundwater Recirculation Pump:* Switch panel operator for water recirculation pump (HS-2) to "off."

6.3   *Mothballing:* If the groundwater recirculation is to be shut down for an extended period of time, see vendor information regarding necessary servicing.

Note: These procedures are presented as originally configured. Some revision may have occurred if equipment and systems were modified during retrofitting.

3.6.2 Non-Routine Operating Procedures

General Emergency

Non-routine operating procedures should be implemented in the event of sudden or unplanned changes such as electrical failure, fire, or severe weather. Characterization of these situations and appropriate actions are provided in Table 2.9 and are the same for ECRS Units 1 and 2.

Table 3.6 Air Injection System Operating Procedures — Unit 2

1.0   Soil Air Injection Startup

  1.1   *Process Connections:* Inspect general condition of process piping. Check process connections and hoses between process equipment skid, soil tank, and ancillary equipment. Close all sample ports.

  1.2   *Air Compressor:* Switch panel operator for the air compressor (HS-6) to "auto." Monitor main air supply pressure (PG-8) until maximum operating pressure is reached (approximately 120 psi).

  1.3   *Injection Air Supply:* Open sparge supply valve BV-3.

2.0   Injection Mode

  2.1   *Continuous Injection:* Switch panel operator for sparge (HS-1) to "on."

  2.2   *Pulsed Injection:* Switch panel operator for sparge (HS-1) to "auto."

3.0   Adjusting Injection Pressure

  3.1   *Increase Pressure:* Increase setpoint pressure at injection regulator (PRV-7) located on gauge panel (CP-2) by turning regulator adjustment knob *slowly* clockwise.

  3.2   *Decrease Pressure:* Decrease pressure at injection regulator (PRV-7) located on gauge panel (CP-2) by turning regulator adjustment knob *slowly* counter-clockwise.

4.0   Adjusting Pulsed Injection Timers

  4.1   *Pulsed Injection Timers:* Air injection timers are software functions in the data recorder (UIR-1, located in CP-1). Adjustments are made through modification of the data recorder configuration.

5.0   Post-Startup Inspection

  5.1   *Air Injection System Check:* check pressure gauges, temperature gauges, and flow meters to ensure that the air injection system is working properly and that an appropriate flow rate is established.

6.0   Normal Shutdown

  6.1   *Injection Pressure:* Set injection pressure to be minimum.
  6.2   *Injection Mode:* Switch panel operator for sparge (HS-1) to "off."
  6.3   *Injection Air Supply:* Close injection supply valve BV-3.
  6.4   *Air Compressor:* Switch panel operator for the air compressor (HS-6) to "off."

Note: These procedures are presented as originally configured. Some revision may have occurred if equipment and systems were modified during retrofitting.

Freeze Protection Plan

The ECRS system is not designed to operate under freezing conditions and has no freeze protection on the tanks, piping, or instrumentation. The procedure detailed in Table 3.9 is designed to prevent or minimize freeze damage from sustained cold temperatures. Water stored in excess of 1,000 gal in the water storage reservoir (T-2) is not expected to freeze during short-term cold temperatures. If temperatures are predicted to remain below 0°C for longer than 24 hr, the procedures outlined in Table 3.9 should be implemented.

Table 3.7 Soil Vapor Extraction System Operating Procedures — Unit 2

1.0   Soil Vapor Extraction Startup

1.1   *Process Connections:* Inspect general condition of process piping. Check process connections and hoses between process equipment skid, soil tank, and ancillary equipment. Close all sample ports.

1.2   *Air Compressor:* Switch panel operator for the air compressor (HS-6) to "auto." Monitor main air supply pressure (PG-8) until maximum operating pressure is reached (approximately 120 psi).

1.3   *SVE Venturi Pump:* Supply motive air and regulation to SVE venturi pump (SP-1). Open motive air supply valve BV-6 (pipe rack). Open SVE regulator supply valve BV-5 (on CP-1).

2.0   Adjusting SVE Vacuum

2.1   *Increase Vacuum:* Increase vacuum at SVE regulator (PRV-4) located on gauge panel (CP-2) by turning regulator adjustment knob *slowly* clockwise.

2.2   *Decrease Vacuum:* Decrease vacuum at SVE regulator (PRV-4) located on gauge panel (CP-2) by turning regulator adjustment knob *slowly* counter-clockwise.

3.0   Post-Startup Inspection

3.1   *SVE Check:* Check actual SVE in ECRS tank by monitoring pressure gauges, temperature gauges, and flow meters to ensure that SVE performance is within expected range.

3.2   *Process Connections:* Inspect the general condition of the process piping. Check process connections and hoses between process equipment skid, soil tank, and ancillary equipment.

4.0   Normal Shutdown

4.1   *Decrease Vacuum:* Set SVE vacuum to minimum.

4.2   *SVE Venturi Pump:* Isolate SVE venturi pump (SP-1). Close SVE regulator supply valve BV-5 (on CP-1). Close motive air supply valve BV-6 (in pipe rack).

4.3   *Air Compressor:* Switch panel operator for the air compressor (HS-6) to "off."

Note: These procedures are presented as originally configured. Some revision may have occurred if equipment and systems were modified during retrofitting.

3.6.3   Waste Handling and Disposal

Waste materials generated during experiments conducted in the ECRS will include the media in the soil tank, used personnel protective equipment (PPE), spent granular activated carbon (GAC), and used filters for air and water. Researchers will determine the appropriate disposal practices based upon the types of contaminants and their concentrations. General guidelines and applicable regulations are outlined below.

Waste Media

The principal investigator is responsible for contacting a disposal facility regarding requirements for waste profiling, manifesting, transportation, disposal, and decontamination of the soil tank. The researcher is responsible for ensuring that disposal is conducted in accordance with regulations of the host facility as well as applicable federal, state, and local regulations.

Table 3.8 System Shutdown Procedures — Unit 2

1.0   Emergency Shutdown

    1.1   Critical Emergency
        i)   injury to personnel
        ii)   fire
        iii)   major equipment malfunction
        iv)   catastrophic release of process vapor or liquid
        v)   sudden weather event
        vi)   emergency event in nearby facility
        vii)   other impending critical situation
    1.2   *Critical System Shutdown:* Engage the emergency stop push-button (PB-1, located on CP-1).
    1.3   *Electrical System Shutdown:* Remove electrical power from system.

2.0   Normal Shutdown

    2.1   *Shut down* air injection (sparge) system.
    2.2   *Shut down* SVE system.
    2.3   *Shut down* groundwater recirculation system.
    2.4   *Shut down Electrical System:* Remove electrical power from system. Lockout instrumentation building power, process skid power, and main system power.
    2.5   *Process Connections:* Close all process connections and sample ports.
    2.6   *Control Panels:* Inspect general condition of control panels (CP-1, CP-2). Adjust all hand-loading regulators to fully unloaded positions (full counter-clockwise). Adjust all panel operators to "off."

Note: These procedures are presented as originally configured. Some revision may have occurred if equipment and systems were modified during retrofitting.

Used PPE and Used Filters

If the principal investigator determines that used PPE and used filters have been contaminated with hazardous waste, the PPE and/or filters should be retained on site in a labeled DOT-certified 55-gal drum and then disposed of in accordance with the regulations of the host facility as well as applicable federal, state and local regulations (RCRA 40 CFR 260).

Water Treatment Residuals

The principal investigator is responsible for providing any treatment capacity required for recirculated water. Treatment residuals, such as GAC, may be returned to the supplier for regeneration or retained for disposal. The supplier should be contacted for additional information regarding waste profiling, manifesting, transportation, and recycling facilities.

3.6.4 Data Recorder Programming

The intelligent data recorder on Unit 2 has been configured to record process variables such as temperature, pressure, and volume flow for the SVE, air injection or sparging, and groundwater subsystems (Table 3.10). Recorded variables are logged to the memory card of the data recorder in ASCII format and are available for download to be stored or manipulated by other computer programs. In addition, several derived parameters are calculated on the basis of recorded variables (i.e., SVE mass flow, injection mass flow, and hydraulic gradient in the soil tank). The data

Table 3.9 Freeze Protection Procedures — Unit 2

1.0   *System Shutdown*

     1.1   Perform normal system shutdown *except* the electrical power that will be needed during this procedure.

2.0   *Drain ECRS Soil Tank*

     2.1   Drain water from ECRS soil tank.

3.0   *Drain Process Equipment Skid*

     3.1   *Drain Water Filters:* Leave vents and drains open.

     3.2   *Low Point Drains:* Provide low-point drains. Disconnect water reservoir return (TP-5). Disconnect water reservoir supply (TP-6).

     3.3   *Drain Process Piping:* Collect water removed from process piping at *all* low-point drains that will be opened. Switch groundwater reservoir panel operator (HS-4) to "on." Open BV-9. Open BV-28.

4.0   *Electrical System Shutdown*

     4.1   Shut down electrical system.

5.0   *Drain Liquid Treatment*

     5.1   Follow manufacturer guidelines for draining liquid treatment equipment. Collect liquid for subsequent disposal.

Note: These procedures are presented as originally configured. Some revision may have occurred if equipment and systems were modified during retrofitting.

recorder printout includes color traces on the chart recorder for selected variables. The principal investigator may add derived or charted variables to those listed in Table 3.10 by modifying the data recorder configuration.

3.6.5 General Maintenance

ECRS Unit 2 is designed and constructed for easy use and low maintenance. To extend the life of the system components, equipment inspections and service should be completed on a regular scheduled basis. A summary of required preventive maintenance tasks for Unit 2 is provided in Table 3.11. Additional information regarding specifics of each piece of equipment can be found in AATDF Report TR-97-5, Appendices B through E. A recommended list of spare parts for anticipated service tasks is shown in Table 3.12.

3.6.6 Filter Elements

The process equipment skid is equipped with four filters (F-1 through F-4) whose elements require replacement at frequencies dependent upon system operating parameters. Higher air and water flow rates and greater concentrations of particulates in the influent are variables that may cause elements to require more frequent replacement. A procedure for changing filter elements is provided in Table 3.13.

Table 3.10 Measured, Derived, and Recorded Process Variables — Unit 2

Subsystem	Instrument Item No.	Variable	Measured or Derived	Recorded?	Traced?
Soil Vapor Extraction	TE-2	Temperature	Measured	Yes	No
	PIT-2	Pressure	Measured	Yes	Yes
	FIT-2	Volume Flow Rate	Measured	Yes	No
		Mass Flow	Derived from SVE temperature, pressure, and volume flow rate	Yes	Yes
Air Injection	TE-1	Temperature	Measured	Yes	No
	PIT-1	Pressure	Measured	Yes	Yes
	FIT-1	Volume Flow Rate	Measured	Yes	No
	PIT-4	Pressure Sensor	Measured	Yes	Yes
		Mass Flow	Derived from injection temperature, pressure, and volume flow rate	Yes	Yes
Groundwater	TE-3	Inlet Temperature	Measured	Yes	No
	PIT-5	Inlet Head	Measured	Yes	Yes
	TE-4	Outlet Temperature	Measured	Yes	No
	PIT-6	Outlet Head	Measured	Yes	Yes
	Fit-3	Outlet Volume Flow Rate	Measured	Yes	Yes
		Gradient in ECRS Tank	Derived from inlet and outlet heads	Yes	No

Notes: Variables shown above represent a basic configuration of the data recorder on the process equipment skid, as provided to the initial researcher. Any modifications to the scheme shown above are the responsibility of the principal investigator.

Additional derived variables may readily be calculated from the recorded variables indicated above. The new derived variables may then be recorded and traced, if desired.

Recorded variables are logged on the memory card of the data recorder (UIR-1) in ASCII format.

Table 3.11 Required Preventive Maintenance Schedule — Unit 2

Frequency	Item No.	Name	Maintenance
As indicated by reduced performance	PRV-1 to 7	Pressure Regulators	Clean and remove foreign matter
	XV-1 to 6	Solenoid Valves	Clean valve and valve strainer
	F-3, 4	Recirculated Water Filters	Change filter elements
	V-1	Air Receiver	Clean air receiver drain-trap sump screen
	F-2	Air Receiver Discharge Filter	Change filter element

continued

Table 3.11 (continued) Required Preventive Maintenance Schedule — Unit 2

Frequency	Item No.	Name	Maintenance
	F-1	Air Compressor Discharge Filter	Change filter element when indicator on filter housing enters red zone
	C-1	Air Compressor	Change filter element on air compressor intake
Daily	C-1	Air Compressor	Check coolant level, air discharge temperature, separator element differential, air and oil filter differential pressures
Monthly	PSV-1	Pressure Safety Valve	Inspect valve inserts for ripples, tears, or nicks; check seating surfaces for debris, abrasion, or pitting; check pallet edges and guideposts for burrs, corrosion, or damage
	SG-1	Sight Glass	Inspect connections for evidence of leakage; inspect glass for clouding, scratching, or blemishing; clean exterior of glass with commercial glass cleaner
	C-1	Air Compressor	Check V-belt tension and temperature sensor
Semi-annually	C-1	Air Compressor	Replace coolant filter and clean cooler cores
Annually	P-3	Metering Pump	Change elastomeric parts
	SP-1	SVE Venturi Pump	Check for wear, corrosion, debris, or scale
	NA	Electrical	Inspect components and connections for wear, looseness, dirt, dust, or corrosion
	C-1	Air Compressor	Clean separator scavenge screen and orifice; inspect starter contactors; replace air filter
Biannually	C-1	Air Compressor	Replace shaft seal and coolant
Upon decommissioning	FIT-1,2	Flow Meter Orifice Plate	Inspect for signs of nicks, scratches, abrasion, or dull edges on inside of orifice

Table 3.12 Required Spare Parts — Unit 2

Number	Name	Parts/Supplies	Vendor
C-1	Air Compressor	Coolant filter element, inlet valve air filter, separator element, coolant, high air temperature switch, O-rings, fuses	Ingersoll-Rand
F-1	Air Compressor Discharge Filter	Filter element	Ingersoll-Rand
F-2 to 4	Water and Air Filters	Filter elements	Process Filtration Products
FIT-1, 2	Integral Orifice	Gasket	Rosemount
FIT-1 to 3	Flow Meters	O-rings, fuses	Endress & Hauser
LT-1	Level Transmitter	O-rings	Endress & Hauser

Table 3.12 (continued) Required Spare Parts — Unit 2

Number	Name	Parts/Supplies	Vendor
P-3	Metering Pump	Gasket, seal rings, valve balls, injection check valve spring	LMI/Milton Roy
PIT-1 to 6	Pressure Transmitter	O-rings, allen screws	Endress & Hauser
BPV-1, PRV-3, 6, 8	Pressure Regulators	O-rings	Flow Solutions
PRV-1, 2, 4, 5, 7	Pressure Regulators	O-rings	Control Air
PSV-1	Pressure Safety Valve	Pressure seat ring	Whessoe Varec
UIR-1	Data Recorder	Z-fold paper cassette (100 division chart)	Eurotherm-Chessell
V-1	Air Receiver	Bowl kit	Arrow Pneumatics, Inc.
XV-1 to 6	Solenoid Valves	Rebuild kit	ASCO

Table 3.13 Filter Change Procedure — Unit 2

1.0   Change Water Recirculation Filter (F-3)

1.1   *Personnel Protective Equipment (PPE):* Put on appropriate PPE.

1.2   *Bypass Filter:* Bypass filter as follows:
  A)   Open filter bypass valve BV-19 *or*
  B)   Open filter F-4 isolation valves:
       Upstream isolation valve BV-17
       Downstream isolation valve BV-18

1.3   *Isolate Filter:* Close filter isolation valves, as follows:
       Downstream isolation valve BV-16
       Upstream isolation valve BV-15

1.4   *Drain Filter Housing:* Place a bucket under the drain valve of the water filter. Open vent at top of water filter, and filter drain valve located on bottom of filter housing. Allow all water to drain from filter housing into the bucket.

1.5   *Change Filter Element:* While supporting filter housing with one hand, loosen retaining ring at top of filter with other hand. When retaining ring is free, carefully remove filter housing by pulling straight down. Remove and retain filter for disposal. Place new filter in housing, replace filter housing, and tighten retaining ring to seat filter housing against filter head.

1.6   *Prime Filter:* Prime the filter with water as follows:
       Close filter drain valve at bottom of filter housing.
       Open upstream isolation valve BV-15.
       Open filter vent until all air is displaced from filter.
       Close filter vent.
       Open downstream isolation valve BV-16.

1.7   *Dispose of Filter Element:* Dispose of collected water, spent filter elements, and PPE in accordance with materials management plan.

2.0   Change Water Recirculation Filter (F-4)

2.1   *Personnel Protective Equipment (PPE):* Put on appropriate PPE.

2.2   *Bypass Filter:* Bypass filter as follows:
  A)   Open filter bypass valve BV-19 *or*
  B)   Open filter F-3 isolation valves:
       Upstream isolation valve BV-15
       Downstream isolation valve BV-16

2.3   *Isolate Filter:* Close filter isolation valves, as follows:
       Downstream isolation valve BV-18
       Upstream isolation valve BV-17

continued

Table 3.13 (continued) Filter Change Procedure — Unit 2

2.4	*Drain Filter Housing:* Place a bucket under the drain valve of the water filter. Open vent at top of water filter and filter drain valve located on bottom of filter housing. Allow all water to drain from filter housing into the bucket.
2.5	*Change Filter Element:* While supporting the filter housing with one hand, loosen retaining ring at top of filter with the other hand. When retaining ring is free, carefully remove filter housing by pulling straight down. Remove and retain filter for disposal. Place new filter in housing and replace filter housing. Tighten retaining ring to seat filter housing against filter head.
2.6	*Prime Filter:* Prime the filter with water as follows: Close filter drain valve at bottom of filter housing. Open upstream isolation valve BV-17. Open filter vent until all air is displaced from filter. Close filter vent. Open downstream isolation valve BV-18.
2.7	*Dispose of Filter Element:* Dispose of collected water, spent filter elements, and PPE in accordance with materials management plan.

Note: These procedures are presented as originally configured. Some revision may have occurred if equipment and systems were modified during retrofitting.

CHAPTER 4

System Modifications — Units 1 and 2

When each ECRS unit was placed into service, operational data were collected and an evaluation of the equipment and system design was conducted. Feedback from the researchers using the ECRS units, field observations, and additional experimental requirements were used to evaluate modifications that could be made to simplify system operation and improve system performance. These modifications were implemented during maintenance activities.

4.1 SYSTEM PERFORMANCE

The following principal performance issues were identified for both Units 1 and 2:

- System electrical controls and interlocks were difficult to use
- Groundwater level control did not provide fine control
- Air injection pressure spikes occurred when injection was initiated

These issues are described in greater detail in the following sections. Descriptions of the system modifications made to Units 1 and 2 to address these issues are also provided.

4.1.1 System Controls

Both units originally employed an integrated data collection and process control system. The concept was to provide a single interface for the user. The control system was designed to monitor and record process parameters and to test the measured values against predetermined interlocks. The user would provide additional input, such as starting or stopping a pump or blower by toggling switches tied into the control system.

As part of the original design, interlocks were implemented to prevent unintended system operation. This had several undesirable side effects including delayed system response to user inputs. On some occasions, the user was prevented from using any portion of the system due to an alarm condition in one portion of the system (i.e., the groundwater pump could not be operated because of a fault in the air injection blower). Also, the interlocks, as designed, proved to be difficult to modify in the field. In some cases, equipment could not be operated in the desired sequence and several of the hard wired interlocks were too restrictive to allow the desired experiments to be conducted easily.

4.1.2 Groundwater Level Control

Groundwater level control was a continuing issue for both units. Unit 1 was designed so that the flow rate through the closed system was adjusted to the desired constant rate. For a given volume of water and a constant flow rate, a stable gradient was to be established across the test cell. The constant flow rate was accomplished through the use of variable speed gear pumps. The discharge pressure of gear pumps is relatively insensitive to moderate changes in flow rate (i.e., they have very steep pump curves). In several cases, the desired gradient required flow rates that corresponded to the upper or lower limit of the pump performance curve resulting in poor flow control. Also, filtration was required upstream of the gear pumps to remove particulate matter entrained from the soil tank. When the filters began to accumulate particulates, the suction pressure available to the pumps began to drop. This further exacerbated the flow control problems.

Unit 2 was designed to control the inlet and outlet pressure (level) at each end of the soil tank, to establish the absolute liquid level in the cell as well as the gradient. The flow rate corresponding to this gradient could then be recorded. The inlet and outlet pressures were controlled by pilot-driven pressure regulators. A centrifugal pump not requiring suction filtration was used to provide a source of water at a pressure and flow rate greater than that required by the soil tank. The excess water was returned to a surge tank to be recirculated to the pressure regulators. This arrangement solved the problems seen in Unit 1 of controlling the absolute liquid level as well as the gradient across the cell. Additionally, it avoided the problems of suction-side filtration. The principal complaint with this arrangement was the lack of precise control of the liquid level. Level control was dependent on the mechanical response of the pilot valves to changes in the liquid level within the soil tank and mechanical tuning of needle valves to adjust the sensitivity of the pilot and speed of response. The tuning procedure was tedious and care had to be exercised to prevent the accidental adjustment of the needle valves. Retuning was also necessary as the ambient environment changed (i.e., summer to winter). Scale buildup in the lines and pilot valves also affected performance.

4.1.3 Air Injection Pressure Spikes

Pressure spikes were noticed in the air injection system each time it was enabled. In each system the on/off pulsing of the injection system was accomplished by opening and closing a solenoid valve between the pressure regulator and the test cell. The pressure regulator was referenced to a pressure signal from the injection point at the test cell on the downstream side of the solenoid valve. While the solenoid valve was closed, no air would be injected into the soil tank and the injection reference pressure would begin to fall off, eventually reaching the SVE setpoint. Air pressure would build up between the pressure regulator and the solenoid valve to the supply line pressure as the regulator tried to drive the injection reference pressure up to the setpoint of the pilot valve. When the solenoid valve opened at the beginning of another injection cycle, the high pressure air accumulated between the pressure regulator and the solenoid valve was released, causing a momentary spike in the sparge pressure until the pilot valve responded and reduced the pressure supplied by the pressure regulator to the test cell. This process repeated itself each time the injection system was cycled.

4.2 SYSTEM MODIFICATIONS

Modifications were made to both Units 1 and 2 to address the performance issues described in the above section. Unit 1 was modified in October, 1997, after demobilization from Shell Westhollow Research Center in Houston, TX and prior to shipment to the U.S. Army Corps of Engineers Waterways Experiment Station in Vicksburg, MS.

Unit 2 was modified in September, 1998, after demobilization from Arizona State University in Phoenix, AZ, and prior to shipment to Rice University in Houston, TX.

4.2.1 ECRS Unit 1

System Controls

The original system controls for Unit 1 were based on the ORS Site-Pro controller. This microprocessor-based controller was commercially available and was intended for autonomous operation at remote sites. Interlock devices such as level switches and pressure switches were wired to the controller, and interlocks were established via permanent programming within the unit and user accessible dip switches.

The Site-Pro was removed from the system controls. The control logic originally contained with the Site-Pro was replaced with relay logic. Most interlocks were removed leaving only those that protected a single piece of equipment (i.e., motor overload relays for the blower and compressor). Additionally, a physical interlock between the SVE and air injection pressure regulators was removed to uncouple the systems and allow independent operation.

Operational interlocks are now primarily institutional instead of hardwired. It is the responsibility of the researcher to ensure that the desired mode of operation will not result in an unsafe condition or damage to the system. For example, when injecting air into the sealed soil tank, the user must establish a vacuum in the soil tank prior to beginning injection or the flexible top can inflate and eventually rupture.

Groundwater Level Control

The scheme of controlling the flow rate to the test cell and allowing the absolute liquid level and gradient to seek equilibrium was retained. Improvements were achieved by revising the groundwater supply system. The original installation was composed of two gear pumps in parallel. In the revised installation, one gear pump was replaced with a centrifugal pump and placed in series with the remaining gear pump. As the centrifugal pump was less sensitive to particulate matter, the filtration was moved from the suction side of the pumps to the discharge side of the centrifugal pump, between the centrifugal pump and the gear pump. This had two effects. First, it allowed pressure filtration, which resulted in greater filter life as more particulate matter could be accumulated before the suction pressure of the gear pump was adversely affected. Second, with the greater suction pressure available and the pumps placed in series, the gear pump was required to add less dynamic head to the flowstream resulting in more stable operation.

Air Injection Pressure Spikes

The issue of injection pressure spikes was resolved by relocating the reference point for the pressure regulator from the test cell to a location in the piping between the pressure regulator and the solenoid valve. In the revised installation, while the solenoid valve was closed the pressure between the regulator and the solenoid valve increased to the regulator setpoint. As the setpoint was reached, the regulator began to close, shutting off the flow of air into the piping between the regulator and the solenoid valve. When the solenoid valve opened, the air began to flow to the soil tank, reducing the pressure at the reference point and causing the regulator to increase the flow of air. A small pressure variation was noted in the revised installation, due to the response characteristics of the pressure regulator, but it was deemed to be insignificant.

4.2.2 ECRS Unit 2

System Controls

The original control system for Unit 2 was based on a Eurotherm Chessel 4250M Process Recorder. This intelligent data recorder was selected for its ability to record process data, perform mathematical functions, and provide user-accessible interlock logic. Data collection and math functions, such as mass flow and gradient calculations, were maintained in the process recorder. System control logic was replaced with relay logic and loop controllers. Most interlocks were removed, leaving only those that protect a single piece of equipment (i.e., low flow switch to protect the groundwater recirculation pump from running dry).

Setpoints for groundwater level control and air injection pressure are now entered into separate loop controllers through a touchpad not requiring knowledge of programming language or format. Each control loop is separate from the others and may be operated independently.

Operational interlocks are now primarily institutional instead of hardwired. It is the responsibility of the researcher to ensure that the desired mode of operation will not result in an unsafe condition or damage to the system. An example situation is injection into a closed soil tank. The user must establish a vacuum in the soil tank prior to beginning air or the flexible top could inflate and eventually rupture.

Groundwater Level Control

Absolute groundwater level control within the test cell was originally accomplished using mechanical pilot valves to drive the groundwater supply and return pressure regulators. Level control was dependent on the mechanical response of the pilot valves to changes in the liquid level within the test cell. In order to increase the precision of the level, the pilot valves were replaced by loop controllers, with proportional, integral, and derivative functions (PID) to measure the actual level against a setpoint, and by current-to-pressure (I/P) relays to drive the pressure regulators based on output signals from the loop controllers. The PID controllers were tuned to provide a fast response with minimal overshoot of the setpoint. This tuning, once complete, required no further adjustment.

Sparge Pressure Spikes

As in the original Unit 1 installation, the original Unit 2 air injection pressure control was referenced to the pressure injection point leading to similar pressure spike problems. In this case, the problem was solved by implementing a PID loop controller to control the setpoint of the pressure regulator. The PID controller was adjusted to provide a fast response with minimal overshoot of the setpoint. An injection timer was used to make a step adjustment to the setpoint of the PID controller at the beginning and end of each injection event. The setpoint was switched from 0 psig to the user-entered setpoint at the beginning of the cycle. At the end of the cycle, it was switched from the user-entered setpoint back to 0 psig. This change in setpoint allowed the pressure reference point to remain at the injection point. It also resulted in a smooth transition between the user-entered injection pressure and the background pressure of the test cell.

CHAPTER 5

ECRS Shipping and Setup

5.1 PACKING, ASSEMBLY, AND DISASSEMBLY

The ECRS modular equipment can be packed onto one air-suspension flatbed trailer for shipping. The soil tank, with water reservoir packed inside, is winched onto the rear of the trailer (Photo 5.1), and the other equipment modules (instrumentation building, process skid, and shelter) are loaded onto the trailer using a forklift or crane and secured with tarps and straps (Photo 5.2). At the delivery point, the equipment modules are removed first, and then the tank is rolled off the trailer and into position at the site using the winch.

Photo 5.1   Soil tank, with water reservoir packed inside, winched onto rear of transport trailer.

When positioning the soil tank at the site, consideration is given to clearance around the rear door on the tank and access to hose attachments. At the end of the project, soil removal is facilitated by opening this door to provide access for a small front end loader. Hoses from the process equipment skid will need to be able to reach the attachments on the front of the soil tank.

The soil tank is then packed with media and instrumentation (Photos 5.3 and 5.4), and the top is sealed to the lip of the tank using caulk and angle-iron bolted through the lip (Photos 5.5 and 5.6). The shelter frame is unpacked and erected (Photos 5.7 and 5.8), and the process equipment skid and water reservoir are positioned near the soil tank and under cover of the shelter (Photo 5.9). The climate-controlled instrumentation building can be positioned outside or under the shelter.

Photo 5.2 ECRS Unit 2 modules packed for shipping (crated shelter, process skid, instrumentation building, and soil tank).

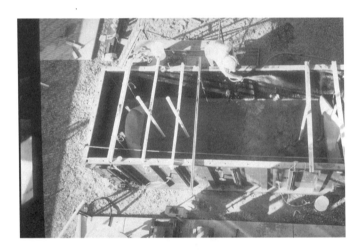

Photo 5.3 Soil tank readied for packing with media and instrumentation.

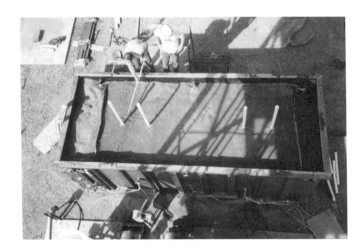

Photo 5.4 Soil tank packed with sandy soil prior to placement of gravel pack and fabric top.

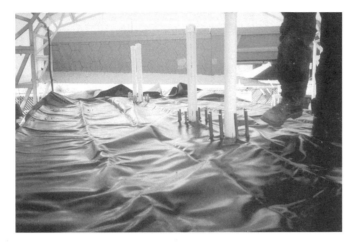

Photo 5.5   Installation of fabric top and manways on soil tank.

Photo 5.6   Edge sealing system of angle-iron, bolts, and caulk for soil tank fabric top.

Photo 5.7   Unpacked shelter frame for ECRS modules.

Photo 5.8   Shelter frame being erected over ECRS equipment.

Photo 5.9   Fabric cover installed on shelter housing ECRS Unit 2 modules.

5.2 SOIL PACKING AND REMOVAL

Soil packing and instrumentation are designed by the researcher. Soil is stored in a staging area near the tank and transported to the tank using a front-end loader (Photo 5.10). Well screens and infiltration galleries can be packed with commercial-grade sands or gravels, also transported into the tank using the front-end loader. The sand can be loaded into the tank in lifts of a designed thickness and compacted using commercial vibrators (Photo 5.11) and water. Sampling lines can be placed on the compacted sand surface and buried beneath successive lifts of soil. Wells can be installed in the compacted soil using hand augers. Sampling lines can be fed to manifolds on the front of the tank or in the tank top (Photo 5.12).

The fully packed soil tank is too heavy to winch onto a trailer. Nonhazardous soil can be removed from it after completion of a project by unlatching the sealed tank door and using a small front-end loader to haul soil out. If the soil contains hazardous chemicals, a portion can be removed to another roll-off box using a larger front-end loader, and then both partially full soil tanks can be winched onto a trailer for hauling to an appropriate disposal facility.

Photo 5.10 Front-end loader delivering sand to pack soil tank.

Photo 5.11 Mechanical compaction of soil during tank packing.

Photo 5.12 Sampling lines from the soil pack fed through manifolds on soil tank.

5.3 RELOCATION COSTS

Costs for equipment transportation and assembly or disassembly include a driver, truck with an air-suspension trailer, and travel plus per diem costs for a three-person crew to set up or decommission the equipment over 3 to 4 days. The logistics of site preparation and utility hookups are managed by the researcher prior to delivery of the ECRS unit. The researcher provides the soil, any instrumentation that is placed in the soil pack, and staff to assist with soil packing. For decommissioning, the researcher removes soil from the tank and rinses the soil tank prior to AATDF personnel disassembling and decontaminating the equipment. The researcher is also responsible for disposal of used soil and wastewater from the decontamination process. The maximum total cost for transportation and assembly or disassembly, to a destination within a 1,000-mile radius of Houston, was about $13,000 including personnel costs.

CHAPTER 6

Test Data and Research Projects

6.1 SYSTEM TESTS

6.1.1   Test Objectives

Unit 1 underwent initial systems tests after the equipment was assembled at Shell and passed a safety inspection. Three tests were performed in the packed soil tank to address the tightness of the system and to characterize the air flow and water flow through the soil pack. System tightness was measured with a helium tracer test. Vacuum pressures generated by the SVE compressor were used to track air flow in the soil, and a potassium bromide tracer was used to map water flow in the soil pack. The protocols for these tests were selected and modified by Jonathan Miller and Ira Dortch at Shell's Westhollow Technology Center (WTC) based upon information provided on tank testing from Oregon Graduate Institute. Miller and Dortch have provided data and a report from the systems tests and research (Appendix C, Figures C1.0 to C18.0). A proposed SF_6 tracer test in groundwater, to map water flow, was omitted because of the potential for the SF_6 to sorb to the Teflon sample tubing in the soil tank.

6.1.2   Leak Testing

Tank tightness tests were performed by using helium as a tracer gas to detect air leaks in the system, as illustrated in Figure 6.1. A negative pressure was maintained in the soil pack during the tests by setting the compressor and blower to operate with a 1.1 psi pressure drop across the tank. The test consisted of adding a known volume of helium (5%) to the air stream that is injected into the upper soil pack at one end of the tank (through one horizontal SVE well), and measuring helium concentrations in air extracted from the lower soil pack at the opposite end of the tank (through one horizontal groundwater well). An exit time distribution graph was plotted, which measured the ratio of the extracted helium concentration to the original injected concentration vs. time. Tank tightness was demonstrated when the ratio of the concentrations reached a value of one.

During the first two tightness tests, the seal on the tank top and rear door gasket leaked air. The seal was improved by mechanical smoothing and filling of the tank lip, placement of a new neoprene gasket with vacuum grease, and use of vacuum grease on the door gasket. The third tightness test was successful, and the exit time distribution graph from that test is included as Figure 6.2 (and Appendix C, Figures C7.0 and C8.0).

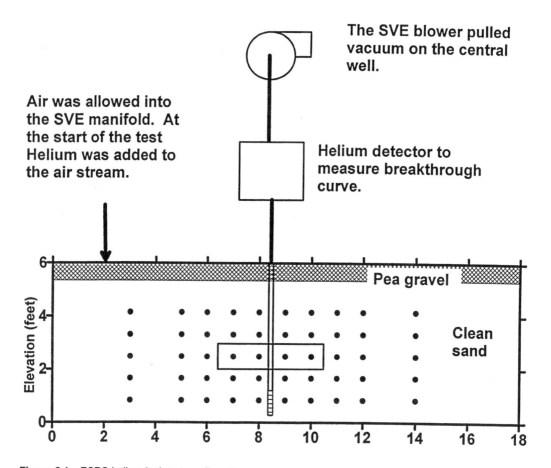

Figure 6.1 ECRS helium leak test configuration.

6.1.3 Flow Tests

Air Flow Test

The air flow test was performed by running the SVE compressor and pulling a vacuum in the two horizontal wells in the upper gravel layer, at each end of the tank, and leaving the central sparge well blower off. Figures 6.3 and 6.4 illustrate the test configuration and the data. Vacuum pressures were measured across the grid of 50 sample frits in the soil over a period of hours (Appendix C, Figures C5.0 and C6.0). The response in the soil pack after 2 hrs of operation is shown in Figure 6.4 (Appendix C, Figure C6.0). The figure shows that a vacuum of approximately 80 in. of water was measured across the upper portion of the soil pack. Vacuum readings reached a minimum of approximately 44 in. of water in the soil surrounding the base of the sparge well, beneath the reduced permeability layer in the center of the soil pack. The vacuum readings indicated that pressures were reduced within and beneath the reduced permeability layer. This demonstrated that the soil pack was responding as designed.

Potassium Bromide Water Flow Test

The potassium bromide test was performed to characterize water flow patterns in the soil pack. A 850 parts per million (ppm) solution of potassium bromide was mixed in one of the water

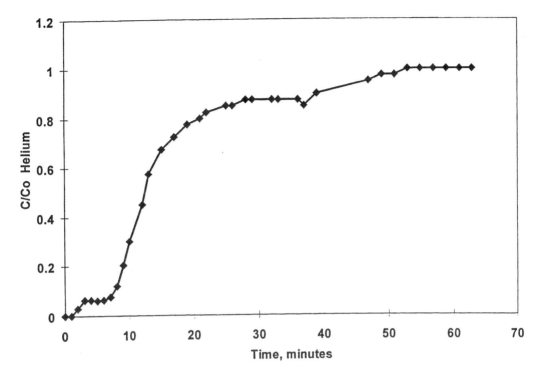

Figure 6.2   ECRS helium leak test results.

reservoirs and pumped at a rate of 0.5 gpm into the base of the soil pack through the horizontal water well at one end of the tank, as shown in Figure 6.5 (Appendix C, Figure C9.0). Water samples were collected from the grid of 50 sampling frits over a period of four days. Figure 6.6 illustrated the progression of the potassium bromide tracer through the soil pack (and Appendix C, Figure C10.0). The tracer data indicated that the water moved out of the injection gallery in a plug flow pattern until it reached the reduced permeability layer in the center of the soil pack. The flow passed through the first two sampling points in the reduced permeability layer, but it was deflected around the rest of the layer and then continued across the tank. This indicated that at least half of the reduced permeability layer was reducing water flow as designed.

6.2 RESEARCH PROJECTS IN UNIT 1

6.2.1 Equilon (Shell Development Co.) — Air Sparging and Oxygen-Releasing Material

Summary

The first ECRS unit was located at Shell Westhollow Technology Center (WTC) from March 1996 through August 1997, for experiments conducted by Jonathan T. Miller and Ira J. Dortch. The unit underwent initial systems tests and safety inspection prior to the research project. Experiments were conducted to evaluate natural attenuation, air sparging, and oxygen-releasing materials (ORMs). A submerged source of trapped hydrocarbon NAPL was created by injecting 2.5 kg of unleaded gasoline into the center of the soil pack in the tank. Steady-state groundwater flow was then established in the soil matrix, and air sparging was initiated. Results of this first test indicated

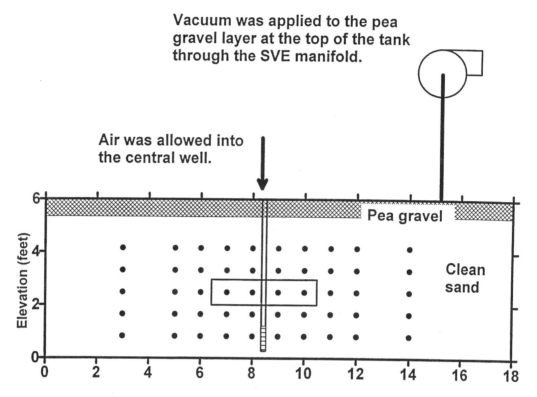

Figure 6.3 ECRS air flow test configuration.

that natural attenuation processes produced large reductions in hydrocarbon mass before air sparging began, which prevented determination of the beneficial effects of the air sparging on mass removal. In a subsequent test, air sparging was effective at increasing levels of dissolved oxygen in the groundwater, but oxygen transfer rates fell off sharply after steady air flow was established, presumably due to development of discrete air channels in the soil matrix. The ORM data compared the oxygen-releasing capacity and duration of commercial and generic ORMs. The results were proprietary within Shell.

Project: Optimization of Air Sparging for Remediation of Soils and Groundwater Impacted by Petroleum Hydrocarbons

The ECRS soil tank was packed with a heterogeneous mixture of compacted sand to study contaminant removal via soil vapor extraction and air sparging. Goals of this experiment were to 1) investigate and characterize a dissolved phase plume emanating from a contaminant source area in the saturated zone; 2) investigate the effectiveness of air sparging in removing residual contaminants under conditions of heterogeneous layering; 3) investigate the effect of alternative remediation measures, including pulsed sparging, soil vapor extraction, and oxygen-releasing materials; and 4) prove-out the ECRS facility and troubleshoot any problems in the design.

Pressure and flow data were collected from several sensors arranged on the equipment and stored automatically using a data logger. In addition, a dedicated gas chromatograph analyzed samples of the off-gas from the tank every hour, and fluid samples were periodically collected through 50 sampling ports installed directly in the soil. These samples were used to measure the spatial distribution of dissolved oxygen and hydrocarbon constituents.

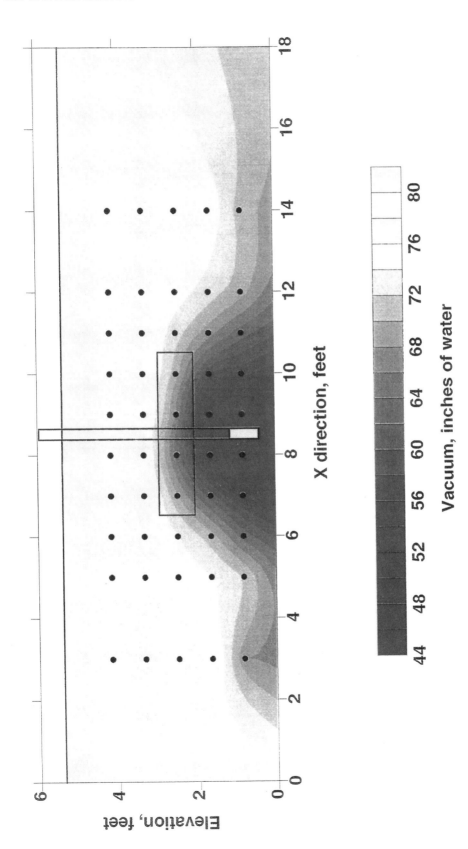

Figure 6.4 ECRS air flow test results.

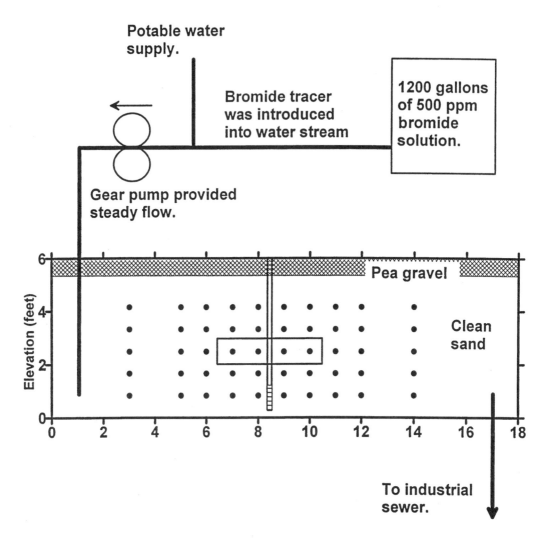

Figure 6.5 ECRS water tracer test configuration.

The tank was packed with clean masonry sand, and a lens of less-permeable bank sand 10 in. thick and 4 ft long was centrally located within the tank as illustrated in Figures 2.1 and 2.2 (and Appendix C, Figures C1.0 to C4.0). The difference in permeability between the two sands was a factor of about 2, representing a fairly mild spatial variability. The intention was to study air sparging beneath heterogeneities that had a small grain size change that was undetected in a typical well-boring log, but significant enough to degrade sparging performance.

After the apparatus was set up, a series of tests were run to verify that the components of the ECRS facility were working properly. In the first test, pressures were measured at the 50 sample ports under conditions of both steady-state air flow (vadose zone) and steady-state water flow (saturated zone). This served to verify that the sampling tubes were not plugged and were labeled correctly. A helium tracer test then indicated that the seal of the tank was good. Helium was injected into the inlet air stream and detected at the outlet as it broke through the soil. All the injected helium was recovered. In a third test, bromide tracer was injected into the inlet water flow under steady-state saturated flow conditions. Samples were drawn every 12 hrs from the 50 sampling ports to track the bromide as it passed through the soil. These tests confirmed

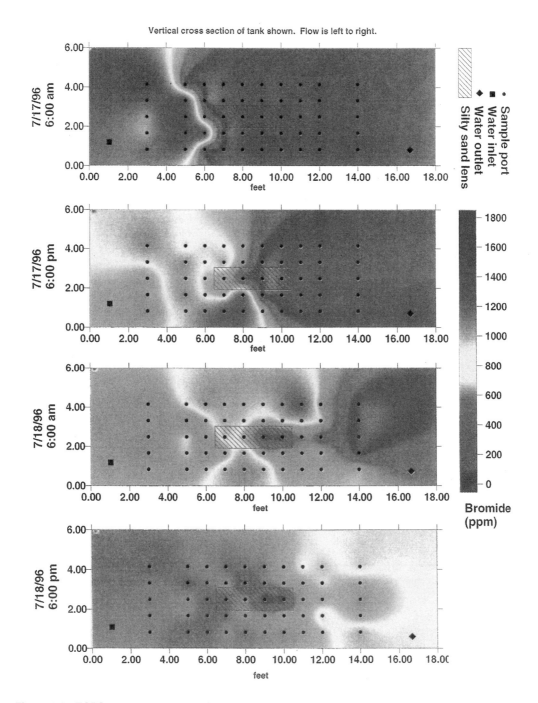

Figure 6.6 ECRS water tracer test results.

that the components of the system were working properly and also characterized flow in the heterogeneous packing.

After the tests were complete, the tank was drained halfway and approximately 2.5 kg of Shell regular unleaded gasoline was allowed to drain by gravity through a special tube. The tube delivered the gasoline onto the capillary fringe at the top of the silty lens near the center of the tank. The gasoline was allowed to settle for two days, then the water table was raised to create a submerged source area of hydrocarbons. A steady-state groundwater flow of approximately 15 ft/day was established in the tank. The water was cleaned with activated carbon before returning to the tank inlet. Figure 2.1 (and Appendix C, Figure C2.0) shows a cross-sectional view of the soil pack, well screens, and sampling array in the soil tank.

Natural Attenuation

The tank configuration during the natural attenuation experiments is shown in Figure 6.7 (Appendix C, Figure C11.0). For this experiment, water samples drawn from the 50 sampling ports and from the tank influent and effluent lines were analyzed for BTEX and MTBE using a gas chromatograph to monitor migration of the dissolved phase plume away from the source area. Figure 6.8 summarizes the natural attenuation data for total BTEX, and Figure 6.9 summarizes the data for BTEX compounds (and Appendix C, Figures C12.0 and C13.0, respectively). The total mass of BTEX was calculated by multiplying the total mass of gasoline by the weight fraction of BTEX:

$$2.495 \text{ kg gas} \times 0.182 \text{ kg BTEX/kg gas} = 0.454 \text{ kg BTEX}$$

The total mass of BTEX removed by the activated carbon was calculated by multiplying the flow-rate time-averaged concentration in the effluent by the duration of the experiment:

$$0.4 \text{ gal/min} \times 3.78 \text{ L/gal} \times 1 \text{ kg water/L water} \times 0.0857\text{E-6 kg BTEX/kg water}$$
$$\times 119 \text{ days} \times 1440 \text{ min/day} = 0.0222 \text{ kg BTEX}$$

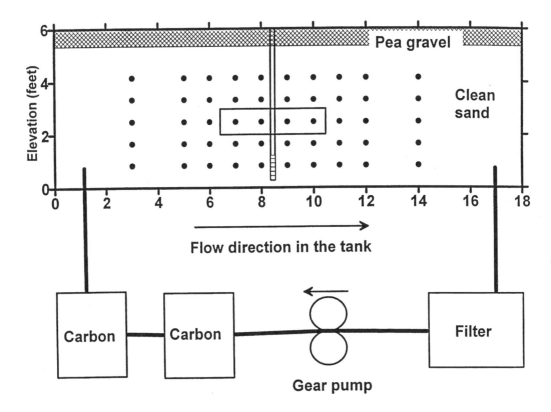

Figure 6.7   ECRS natural attenuation experiment configuration.

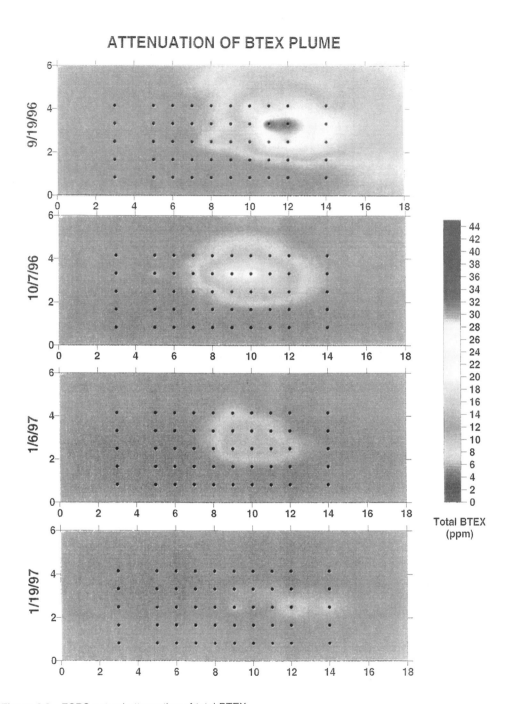

Figure 6.8 ECRS natural attenuation of total BTEX.

This value represented 5% of the injected BTEX mass. There was some uncertainty in this measurement because of initial interruption in the steady-state flow by the controller unit on the process equipment skid. The averaged concentration used in this calculation was based upon three samples collected from the effluent line.

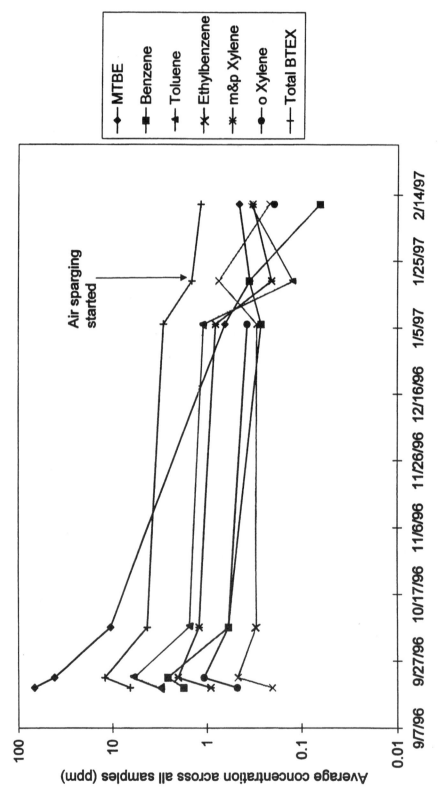

Figure 6.9   ECRS natural attenuation of BTEX compounds.

The total mass of MTBE was calculated in a similar manner to BTEX:

$$2.495 \text{ kg gas} \times 0.11 \text{ kg MTBE/kg gas} = 0.275 \text{ kg MTBE}$$

The total mass of MTBE removed from the effluent stream by the granular activated carbon was calculated using the following formula:

$$0.4 \text{ gal/min} \times 3.78 \text{ L/gal} \times 1 \text{ kg water/L water} \times 0.949\text{E-6 kg MTBE/kg water}$$
$$\times 199 \text{ days} \times 1440 \text{ min/day} = 0.256 \text{ kg MTBE}$$

This represented 89% of the injected MTBE mass, and is discussed in the following sections. Over the course of the four-month experiment, the reduction in BTEX concentration was considerable. Average BTEX and MTBE concentration levels are plotted in Figure 6.9 (and Appendix C, Figure C 13.0) using a log-scale vertical axis. Assuming that these concentrations follow an exponential decay law, the decay constant was found:

$$K(\%/\text{day}) = \ln(C_{final}/C_{initial})/\text{elapsed days}$$

where $C_{initial}$ = concentration at the beginning of the test
C_{final} = concentration at the end of the test period

The following decay rates were calculated using the following percentages:

	BTEX	**MTBE**
Before Sparging 10/22/96 to 1/19/97	1.8%/day	4%/day
During Sparging 1/19/97 to 2/11/97	0.9%/day	−1.1%/day

There were significant reductions in BTEX and MTBE mass from natural attenuation before air sparging began. The apparent decay rates of 1.8%/day for BTEX and 4%/day for MTBE were much higher than typical rates reported from field sites. This is probably due to the rapid ground-water flow rate through the tank and the fresh gasoline spill. Lighter components did not have time to volatilize, and heavier components did not have time to sorb onto the soil particles. In the case of MTBE, decay was not related to biodegradation, but resulted from transportation of the mass in the dissolved phase out of the tank, as shown by the mass balance for MTBE. For BTEX, the apparent decay rate was slower due to the lower solubility of BTEX, and biodegradation was the dominant mechanism for mass reduction.

Air Sparging

An experiment was also run to evaluate the effectiveness of air sparging at delivering dissolved oxygen to the saturated zone for speeding removal of BTEX. For this test the tank was saturated with groundwater and a steady flow of 15 ft/day was established. Soil vapor extraction was simulated in the matrix by pulling a vacuum of approximately 5 in. of water on the pea gravel layer at the top of the soil pack. Air sparging was initiated by injecting compressed air into the central well inside the tank. The experimental design is shown in cross section in Figure 6.10 (and Appendix C, Figure C14.0).

Dissolved oxygen measurements were taken from water samples drawn through the 50 sampling ports both before and during the air sparging. The air pressure and flow rate at the time of sparging initiation are shown in Appendix C, Figure C15.0. Both air pressure and flow rate dropped immediately after sparging was started, pressure showing the largest drop. This effect is well known from air sparging field studies and thought to result from development of stable air channels in the soil

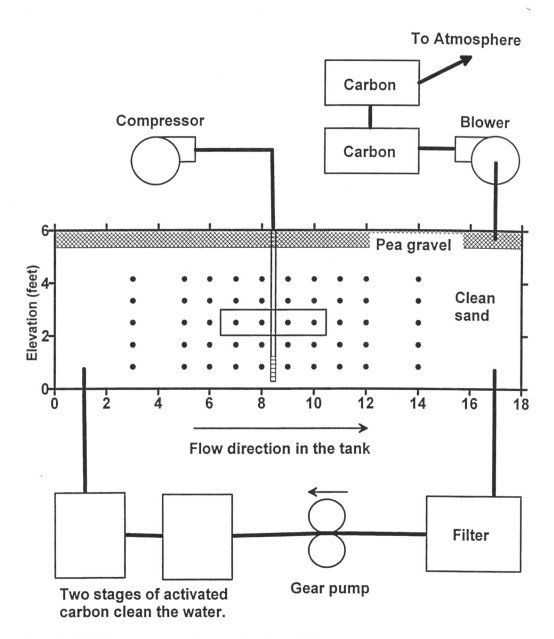

Figure 6.10 ECRS air sparging equipment configuration (Equilon).

matrix. When the channels become established, the pressure gradient needed to drive air sparging is reduced. Dissolved oxygen data, discussed below, also suggest channel formation during sparging.

Changes in water level in the tank as a result of air sparging are shown in Appendix C, Figure C16.0, which illustrates that the water level rose and remained elevated during sparging. This mounding of the water table has been observed during field tests involving air sparging (Johnson et al., 1993) and results from displacement of water by the injected air. Mounding has a relatively short duration in field studies, in the range of minutes to hours, before the water table returns to its original position. The mounded water flows laterally in a natural aquifer, dissipating the elevation difference. This later flow was restricted in the ECRS soil tank and the mound cannot dissipate.

Assuming the total volume of liquid in the ECRS system was constant, the mound should have indicated the total volume of air that was added to the soil matrix during steady sparging. Assuming the water table rose 5 in uniformly in the tank, the volume of that water represented

$$V = 5 \text{ in} \times 7 \text{ ft} \times 18 \text{ ft} \times (1 \text{ ft/12 in}) = 52.5 \text{ ft}^3$$

The total saturated volume in the tank was about 500 ft^3 (assuming 7 ft $\times$ 18 ft $\times$ 4 ft saturated thickness). With a porosity of about 30% in the soil pack, 150 ft^3 of pore space was left to be occupied by injected air. Therefore, the estimate of injected volume based on mounding suggested that fully one third of the available pore volume was occupied by injected air during sparging. That result was much higher than recent simulations that predicted a maximum air saturation of 0.3 to 0.4 near the injection point and a steep dropoff away from this zone, suggesting that at the point where the air enters the soil, it displaced roughly only one third of the water present.

The first experiment on natural attenuation was terminated when sparging began to allow a clear evaluation of sparging against the background decay rate. Appendix C, Figures C12.0 and C13.0, show BTEX removal during the period of natural attenuation and then after sparging. These figures suggest only a marginal increase in removal rate during sparging.

Dissolved oxygen (DO) levels before and during sparging are shown in Figure 6.11 (and Appendix C, Figure C17.0). After 20 continuous hours of sparging, oxygen levels near the sparge well rose to 8.5 ppm. The distribution of DO in and above the central silty lens suggested that sparge air was able to penetrate and pass through this finer grain zone even though water flow could not, based on the potassium bromide tracer test. The air penetration of this zone may have been facilitated by the formation of stable air channels. After 24 days of steady sparging, dissolved oxygen levels had returned to the background level, which also supported the formation of stable air pathways through the soil matrix. Formation of these channels would reduce the rate of oxygen transfer into the groundwater by reducing the total interfacial area between the gas and the water in the matrix.

The tank configuration for the oxygen-reducing material is shown in Figure 6.12 (and Appendix C, Figure C18.0).

6.2.2 Waterways Experiment Station, USAE — Degradation of Munitions Compounds

ECRS Unit 1 was relocated to the Army Corps of Engineers Waterways Experiment Station, Vicksburg, MS in January 1998. The unit has been assigned to the Corps of Engineers to be used for pilot-scale testing of remediation technologies and also for on-site treatability tests. The first project scheduled for the unit involves research on degradation of dissolved munitions compounds in groundwater.

6.3 RESEARCH PROJECTS IN UNIT 2

6.3.1 Arizona State University — Air Sparging

Summary

The second ECRS unit was located at Arizona State University, Tempe, AZ, from February 1997 through August 1998, for air sparging experiments conducted by Professor Paul C. Johnson and Cristin Bruce. The unit underwent initial systems tests and safety inspection by personnel from GSI and Protec after assembly at the site. The ECRS tank studies will provide a valuable link between the two-dimensional physical model studies and field-scale experiments. It will also

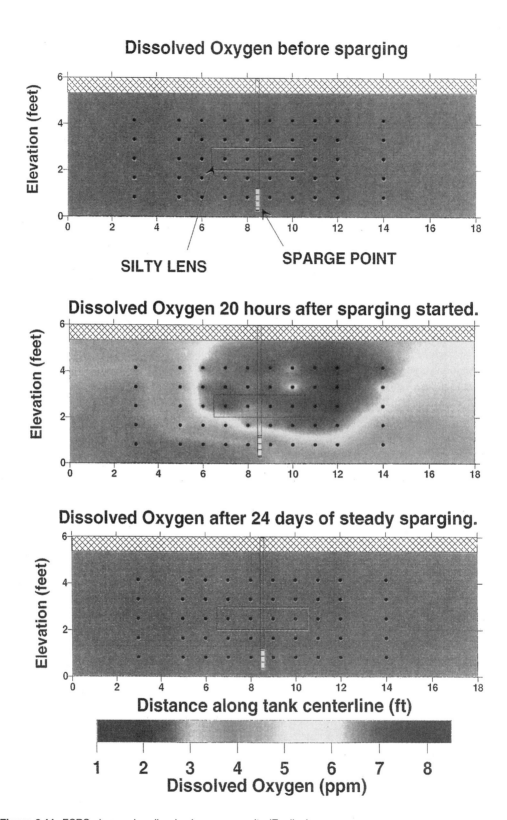

Figure 6.11 ECRS air sparging dissolved oxygen results (Equilon).

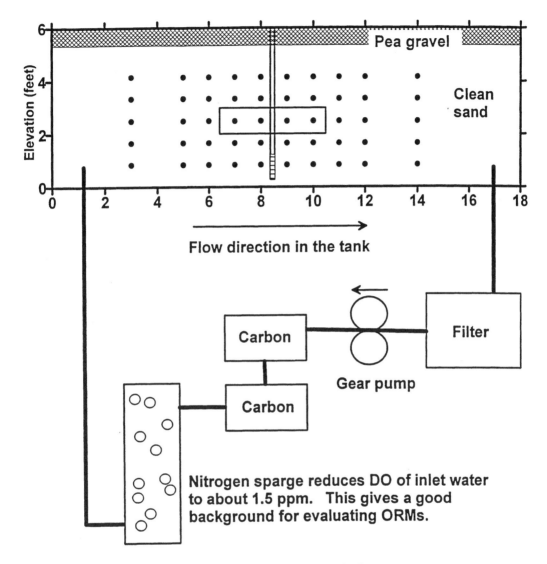

Figure 6.12 ECRS oxygen-releasing material test configuration (Equilon).

provide valuable data for in assessing how the scale of an experiment can affect the results that are observed. The studies were still in progress during preparation of this manuscript. The researchers' latest report to AATDF is included to describe another application for ECRS.

Project

In situ air sparging generally involves the injection of air into an aquifer in order to volatilize and biodegrade contaminants trapped beneath the water table or within the capillary zone. Applications include the treatment of immiscible-phase hydrocarbon source zones and dissolved contaminant plumes and also the use as a barrier to dissolved contaminant plume migration. While air is usually the injected gas, others may be mixed with the injection air stream to promote biodegradation; for example, butane might be blended with the air to promote cometabolic degradation of a chlorinated solvent. *In situ* air sparging has traditionally been used in conjunction with soil vapor extraction, where the chemical vapors liberated by the injected air are collected and treated by the soil vapor extraction system.

Reports from field applications indicate that *in situ* air sparging has been very effective at some sites while less effective at others. It is difficult, however, at this time to determine if these responses are a result of the different site conditions, like hydrogeologic properties and contaminant type, or are due to varied design practices and operating conditions. The mechanisms responsible for the removal are still not clearly understood, and this prevents the development of both conceptual and predictive models for air sparging performance.

The U.S. Air Force and the SERDP program are currently sponsoring research to better understand the performance of *in situ* air sparging systems. In this work, both field-scale studies and physical model experiments are being conducted. The field studies are focusing on the characterization of air distributions within an aquifer at a petroleum fuel release site, while the physical model studies are being conducted to better understand the mechanisms of contaminant removal, such as volatilization and degradation, and how they are influenced by changes in process and hydrogeologic conditions.

The physical model studies are being conducted at Arizona State University and Oregon Graduate Institute (OGI). Specifically, the studies simulate treatment of immiscible-phase hydrocarbon source zones and dissolved phase hydrocarbon plumes. These experiments are designed to identify how changes in air injection rate, air injection pulsing, chemical type, geology, and chemical location affect removal achieved by *in situ* air sparging. The goal is to identify qualitative trends in behavior, rather than to try to quantitatively extrapolate laboratory-scale results to field-scale conditions.

Studies are conducted at various scales in physical models. The smallest physical model is a two-dimensional 8 × 4 ft model aquifer that is 2 in. wide. The intermediate size model is a large-scale, two-dimensional model at OGI, 8 × 30 ft and 1 ft wide, that tips to simulate sloped bedding planes. The largest physical model is the pilot-scale AATDF ECRS Unit 2. The soil tank is 18 × 7 × 6 ft wide.

The experimental plan includes a wide range of scoping studies to be conducted first in the smallest scale physical model. A smaller subset of experiments will then be conducted in the larger-scale physical models to verify that trends observed at smaller scales in two dimensions extrapolate to larger scales and three dimensions. The ECRS tank studies will provide a necessary link between the two-dimensional physical model studies and field-scale experiments. They will also provide valuable data for assessing how the scale of an experiment can affect the results that are observed.

The ECRS soil tank shown schematically in Figure 6.13 is configured so that simulations of treatment in both layered and near-homogeneous settings can be conducted. Contaminants will be introduced either as immiscible-phase liquids through tubes embedded in the soil, or as dissolved-phase contaminants through the recirculating water stream. Inlet and effluent streams will be monitored with an online gas chromatograph, and mass balance accuracy will be assessed using conservative tracer compounds (SF6, He, Br⁻). SF6 was not used in Unit 1 at Shell because of its potential to sorb to the Teflon tubing that was placed in the soil pack. It could be used in the Arizona State University project in Unit 2 because stainless steel tubing was used in the soil pack.

The system at Arizona State University simulates the effect of various remedial technologies on a contaminated shallow water-bearing unit through the use of a 27-yd^3 tank filled with porous media (mortar sand, pea-gravel, and screened fill). For source zone experiments, contaminants are added into the soil through screened injection ports. For the dissolved phase experiments, contaminants are added by a nozzle in the water injection line. Contaminants are removed by soil vapor extraction and water recirculation.

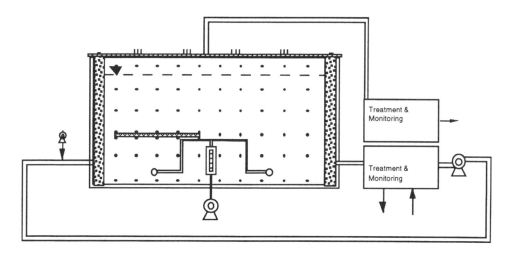

Figure 6.13 ECRS air sparging configuration (Arizona State University).

The goals of the project at Arizona State University are as follows:

- To evaluate the current three-dimensional physical model configuration for its ability to simulate an idealized subsurface
- To demonstrate the contaminant removal efficiencies for *in situ* air sparging (IAS) under different geologic settings

The key components of the test system are a tank connected to a compressor that drives both the sparging and extraction systems, a water circulation system, and a data-logging system. The ECRS tank is filled with approximately 30 tons of mortar sand, and layered on the sides and top with pea gravel. There is a 6-in. layer of screened fill-dirt over half the tank, to demonstrate the effect of nonuniform geologic settings on contaminant removal efficiencies. The compressor can pump ambient air into the tank through any of the three sparging lines connected to the tank (Figure 6.14).

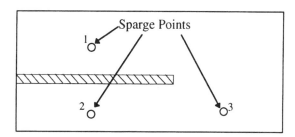

Figure 6.14 Soil sparging setup — Unit 2 (Arizona State University).

Contaminant removal is effected via extraction of air or water contained within the pore spaces of the media filling the tank. The SVE system is fitted with an inline moisture separator, a vacuum relief valve, and an air filter. The water recirculation system is run by gear pumps and fitted with a 1500-gal reservoir connected to a series of coarse and fine filters.

Demonstration Results

Results of initial experiments to determine the mass balance that could be achieved using the current system design were favorable. Over 90% recovery was reproducibly achieved for vapor extraction of gasoline injected into the tank over sparge point #3 (Figure 6.14). Further experiments will examine the ability of IAS to remediate a test mixture of fuel-related compounds.

For source zone experiments:

- Steady air sparging at 5 and 20 scfm for homogeneous porous media
- Pulsed air sparging at 5 and 20 scfm for homogeneous porous media
- Steady air sparging at 5 and 20 scfm for non-homogeneous porous media
- Pulsed air sparging at 5 and 20 scfm for non-homogeneous porous media

For dissolved phase experiments:

- Steady and pulsed sparging at 5 and 20 scfm

Conclusions/Recommendations

Good mass balances from initial tests indicate satisfactory performance of the three-dimensional physical model. Further experimentation will allow determination of optimal *in situ* air sparging design information for the remediation of fuel-related contaminants.

6.3.2   Rice University — Surfactant Biodegradation

Summary

ECRS Unit 2 was shipped from Arizona State University, Tempe, AZ to Protec, Dickinson, TX, in August 1998. There, it underwent maintenance and retrofitting before being assembled at Rice University for experiments, conducted by Professor Joe Hughes and Nelson Neale. The assembly, initial systems tests, and safety inspections will be performed by personnel from Protec. The studies were still in progress during preparation of this book. The researchers' latest report to AATDF is included here to provide another application for ECRS. This one involves surfactant biodegradation and oxygen diffusion in nearly field-scale setting. Preliminary laboratory studies are needed to determine the appropriate conditions to simulate in the ECRS.

Project Methodology and Experimental Design

The project includes two laboratory studies and one nearly field-scale experiment using the ECRS tank. The laboratory work involves oxygen diffusion and surfactant biodegradation studies. All gas analyses will be conducted using a gas chromatograph (Gow-Mac Series 600, Bethlehem, PA). A respirometer (Micro-Oxymax, Columbus Instruments, Columbus, OH) will measure rates of O_2 uptake and CO_2 evolution as well as cumulative values over time for each component. For any ^{14}C studies undertaken, sample analysis will be completed using a Beckman LS6500 scintillation counter (Beckman Instruments, Fullerton, CA). Colorimetric analysis to monitor for surfactant biodegradation will be implemented on a UV-VIS spectrophotometer (Ciba-Corning, Oberlin, OH).

A. Oxygen Diffusion Column Studies — A series of diffusion tests will be run on various soil types to simulate the mass transfer of O_2 through the subsurface matrix and into a contaminated aquifer. The information from this study will be used to determine the role of re-aeration in the biodegradation of subsurface organic contaminants.

Soil Selection — In order to develop a predictive tool for the re-aeration of contaminated aquifers, it is necessary to understand the typical hydrogeologic settings for a variety of stratigraphies that may be found throughout the U.S. In the mid-1980s, the EPA developed an aquifer pollution vulnerability index and ranking system based on the various hydrogeologic settings in the U.S. (Aller, et al., 1987). Based on information from this DRASTIC model, unsaturated zones of aquifers may be classified or described by one of 11 soil types (Table 6.1). These soils will form the basic test matrix for the diffusion studies, excluding gravel and thin or absent categories.

Table 6.1 DRASTIC Soil Type Settings

Soil Type	Description
Nonshrinking and nonaggregated clay	Illitic or Kaolinitic clays with little expansion from water addition
Clay loam	15 to 55% silt, 27 to 40% clay, and 20 to 45% sand
Muck	Fine, well-decomposed organic material. For organic materials, contains at least plant fiber
Silty loam	50 to 85% silt, 12 to 27% clay, and 0 to 50% sand
Loam	25 to 50% silt, 7 to 27% clay, and 0 to 50% sand
Sandy loam	0 to 50% silt, 0 to 20% clay, and 15 to 50% sand
Shrinking and/or aggregated clay	Montmorillonite clays or smectites that swell with water addition
Peat	Soil with undecomposed or partially decomposed, but identifiable, plant material
Sand	Angular or rounded particles from 1/16 mm to 2 mm
Gravel	Particles larger than 2 mm
Thin or absent	No soil cover

Column Design — The diffusion studies will be carried out in a Plexiglas column. The column will measure 1 ft in height by 4 in. in diameter. A wire mesh cloth will be mounted into the column approximately 2 in. from the base and will serve as a barrier between the solid and liquid phases. Two 1/4-in. gaseous sampling ports will be fixed to the side of the column. The first port will be placed approximately 1 in. from the top of the column while the second port will be placed approximately 2 in. from the base near the soil-liquid interface. A 3/8-in. inlet valve will be placed approximately 1 in. from the base and will be used to fill the liquid reservoir.

Experimental Setup and Test Design — Diffusion tests will be conducted on each of the test soils under varying moisture conditions. Specifically, diffusion will be measured at 1/2 and 1/4 field capacity moisture contents. A contaminated aquifer scenario will be simulated in the column by filling the liquid reservoir below the wire mesh cloth with DI water containing sodium sulfite (O_2 scavenger). Gravimetric moisture content of the soil will be measured before and after each test run to study the possible effects of drying.

Determination of O_2 Flux — The flux (J) of gases through soils may be determined by both direct and indirect methods.

Direct flux may be calculated by the following:

$$J_{O_2} = V/A_{cs} \times dC/dt$$

V	=	headspace volume at top of column [cm^3]
A_{cs}	=	cross-sectional area [cm^2]
C	=	O_2 concentration [M/cm^3]
t	=	time [T]
M	=	moles

When O_2 diffusion has reached steady state, which varies depending on soil media and conditions, the column may be sealed with an air-tight top, and the change in O_2 concentration in the headspace with respect to time may be determined using GC analysis. Alternatively, dC/dt may be continuously monitored using a respirometer attached to the column.

Indirect flux may be determined by:

$$J_{O_2} = D_s \times dC/dx$$

D_s = effective diffusion coefficient [cm^2/T]
x = length [cm]

At steady state, O_2 measurement may be made at various depths in the soil column to develop a concentration profile with depth. Flux may then be determined by multiplying this gradient by an estimated D_s. Both direct and indirect flux measurements will be made for each of the diffusion test runs.

Direct Flux Measurement Using Respirometer — Theoretically, dC/dt may be continuously monitored using a respirometer. The Micro-Oxymax respirometer is a closed-circuit system that monitors changes in both O_2 and CO_2 gaseous concentrations using sensors contained within the instrument.

A preliminary comparison of direct flux measurement using the respirometer and syringe sampling and subsequent GC analysis will be conducted to verify accurate calculation of dC/dt by the respirometer.

B. Surfactant Biodegradation Microcosm Studies

B. Surfactant Biodegradation Microcosm Studies — Microcosm studies using a respirometer will be performed to simulate the biodegradation of selected surfactants with re-aeration as the only source of O_2 renewal. Information from this study will be used to determine the fate of residual surfactants in aquifer systems that have resulted from remediation flushing operations.

Surfactant Selection — Three surfactants have been selected for evaluation in this study. A readily mineralizable C_{12} LAS will be used as a reference surfactant for comparative purposes. The remaining two surfactants have been used in field-scale surfactant flushing remediation operations. Dowfax 8390, an anionic n-hexadecyl diphenyloxide disulfonate (DPDS), has been used by researchers at Oklahoma University in field demonstrations at Traverse City, MI (Knox, et. al., 1997). MA-80I, an anionic dihexyl sulfosuccinate, has been used in demonstration studies at Hill Air Force Base, UT, under the direction of Rice University researchers (Hirasaki, et al., 1997). Dowfax 8390 was tested at a concentration of 3.6% (wt.) in the field while MA-80I was evaluated at a 4% concentration.

Inoculum — A microbial seed taken from Dickinson Bayou, TX, sediment will be used for the biodegradation studies. This particular mixed culture has been acclimated to PAH as a sole carbon source. A separate enrichment culture will be set up and maintained for each surfactant in this study.

Experimental Setup and Test Design — Biodegradability of each of the surfactants will be evaluated at three concentrations. Dowfax 8390 will be tested at concentrations of 2%, 3.6%, and 6%, while MA-80I and C_{12} LAS will each be tested at concentrations of 2%, 4%, and 6%. Biodegradation of the surfactants will be studied under optimal conditions (excess nutrients, O_2, and contact) in the absence of soil and under less than optimal conditions, as may be found in subsurface environments. A packed column will be initially flushed with selected surfactants followed by a water rinse to simulate post-flushing conditions. The soil will then be removed from the column and homogenized for use in the microcosm studies.

Determination of Biodegradability — Biodegradation under optimum conditions will be determined by monitoring respirometric O_2 uptake and CO_2 evolution levels. Primary biodegradation will be determined by removing small aqueous samples from the microcosm and measuring for the loss of parent compound using a methylene blue active substance (MBAS) test. Using this information, biodegradation rates will be established for each surfactant and concentration. These rates will serve as a baseline for the maximum amount of biodegradation that may occur under optimal conditions.

Biodegradation under less than optimal conditions (O_2-limiting conditions with overlying porous media) may be more difficult to monitor based on CO_2 evolution. CO_2 has a tendency to remain near the water interface and may not fully transfer back through overlying porous media into the headspace. Thus, only O_2 uptake and reaction stoichiometry may be used for quantitative analysis of biodegradation. Primary biodegradation will again be monitored using MBAS techniques.

C. ECRS Tank Experiment — The ECRS tank experiment combines re-aeration and surfactant biodegradation in a nearly field-scale setting. A single experiment will be conducted to correlate O_2 flux through the unsaturated zone to the attenuation of subsurface surfactant contaminants.

Soil and Surfactant Selection — Both the soil and the surfactant selected for use in this phase of work will be based on the results from the two laboratory studies. Most likely, a coarse-grained sand that does not impede O_2 diffusion and a readily biodegradable surfactant will be selected.

Experimental Setup and Test Design — The ECRS soil tank will be initially packed with the selected soil, and a series of O_2 sensors will be embedded in the soil during packing to monitor for O_2 diffusion. The selected surfactant will be added to the recirculating water reservoir at an appropriate concentration. The water will then be circulated through the base of the tank and the system will be monitored for surfactant biodegradation by MBAS techniques. A carbon balance will be completed on the system.

System Analysis — Results on O_2 diffusion and surfactant biodegradation from the ECRS tank experiment will be compared to the results from the two laboratory studies to determine if the bench-scale results accurately reflected the processes taking place in a nearly field-scale setting.

CHAPTER 7

Findings and Conclusions

7.1 SUMMARY

Two modular remediation testing systems were conceived, designed, constructed, shipped, installed, operated, and monitored. The system design allows cost-effective, quantitative evaluation of a range of technologies in a number of site-specific scenarios.

Design Conception

The ECRS was proposed as a site with containerized cells for testing of technologies by releasing contaminants or chemical amendments into soils. To better define the research niche for ECRS, the AATDF staff gathered data from many existing and proposed technology testing facilities. An ECRS advisory committee was also assembled with representatives from some of those facilities to provide further guidance on the design. Due consideration was then given to data gathered on costs, potential locations, regulatory review and requirements, research uses, and projected schedules for site construction and technology testing. It was decided that construction, regulatory permitting, and maintenance of an in-ground test release site would be costly in terms of both time and funding. The preferred option was to build a portable, pilot-scale testing unit that could be shipped directly to researchers at their home location or a remediation site. This option was also expected to appeal to researchers because it could eliminate their travel costs to distant field sites, allow use of their full research staff, and provide for a longer period of on-site monitoring for their tests.

Design Engineering

The guiding criteria for ECRS design were met. The ECRS was composed of modules that can be shipped to researchers' locations or remediation sites, and the system sealed tightly to facilitate calculation of mass balances. The ECRS provided a cost-effective means of testing the efficacy of remediation technologies or processes at a pilot-scale and had the capability to maintain the soil pack under vadose or ground-water flow conditions. Equipment was fabricated using off-the-shelf components for easy assembly. The researcher was able to pack the soil tank to simulate a variety of subsurface conditions. The equipment and instrumentation accompanying the ECRS would be able to withstand a variety of contaminants, and all instrumentation and equipment were rugged enough to withstand shipping under normal conditions. All the modular equipment for each unit was designed to fit on one flatbed trailer.

AATDF contracted with GSI to assist in preparation of the detailed design package for ECRS Unit 1. For Unit 2, the design was re-evaluated, based upon the performance of Unit 1 and discussion with members of the ECRS Advisory Committee. GSI prepared the revised design package for Unit 2.

Fabrication

Two ECRS units were fabricated. For Unit 1, instrumentation and equipment were assembled at the GSI warehouse in Houston, TX. The soil tank was manufactured and modified by Galbreath, Mansfield, TX. The instrumentation building was fabricated by Bebco, Texas City, TX. The process equipment skid was fabricated at the GSI warehouse from ORS Environmental Systems, Greenville, NH components. GSI also fabricated the manifolds and harnesses for meters and gauges on the front of the tank. This assembly occurred at the warehouse from December 1995 to February 1996.

For Unit 2, GSI prepared the revised design package and a bid package under the guidance of AATDF. Based upon review of bids by potential contractors, Protec was selected as the contractor for Unit 2 fabrication. Unit 2 was assembled at Protec's facility in Dickinson, TX from September 1996 to January 1997.

Shipping and Installation

The ECRS modular equipment was packed onto one air-suspension flatbed trailer for shipping to the site. The soil tank, with water reservoir packed inside, was winched onto the rear of the trailer, and the other equipment modules (instrumentation building, process skid, and shelter) were loaded onto the trailer using a forklift or crane and secured with tarps and straps. At the delivery point, the equipment modules were removed first, and then the winch was used to roll the tank off the trailer and into position.

Operation and Monitoring

Normal operating procedures for ECRS Units 1 and 2 were implemented after installation of the equipment, packing of permeable media and placement of instrumentation in the soil tank, sealing of the soil tank top, connection of electrical service, installation of piping and hoses, positioning of air and water treatment units, and completion of a safety inspection. Startup operations included analytical tests to evaluate the performance of the air injection and SVE systems and the tightness of the sealed soil tank. Regular operations involved setting and monitoring equipment and gauges for three subsystems, including the water circulation system, the air injection or sparging system, and the SVE system. The equipment, controls, and gauges for these systems were located on the process equipment skid for the convenience of the researcher and staff.

Operation and monitoring activities are described in a three-volume operation and maintenance manual that accompanies each unit. General setup, operations, and safety are also described in a video.

7.2 MODIFICATIONS IN UNIT 2 DESIGN

Based upon the performance of Unit 1 during the air sparging research at Shell Westhollow and later quantitative system testing by GSI, it was determined that the system performed to design specifications. After completing the post-performance review, the AATDF solicited comments and suggestions for changes in design or performance from the Shell researchers and advisors, the GSI fabrication team headed by Thomas Reeves, and other ECRS advisors, including Paul Johnson. A summary of major suggested design modifications that were implemented in Unit 2 included a sight glass column, centrifugal pump, relocation of the water filters, screw-type air compressor removal of the SitePro-SpargePro Controller System, one gauge panel on the process equipment skid, humidity control in the instrumentation building, a portable GC, a two-ply urethane fabric top, 1,500-gal opaque water reservoir, air accumulator, supply-air tank on process equipment skid, and portable 50-gal chemical mixing tank.

7.3 GENERAL ATTRIBUTES OF ECRS

The ECRS is a unique, modular testing environment. The major attributes of this system that distinguish it from past test facilities were demonstrated to be the following:

Portable
- containerized testing unit shipped to the researcher's location and set up at modest cost
- can be shipped to a remediation site for treatability tests

Tightly sealed
- system tightly sealed to facilitate mass balance determination

Pilot-scale facility
- intermediate-size test (pilot-scale) between bench-scale and full field-scale
- tank size suited for projects of 1 to 1.5 years duration

Flexible design
- includes SVE, air injection, and ground-water systems
- soil tank size facilitates a range of packing conditions to simulate subsurface environments
- can simulate vadose or aquifer conditions
- can be packed with contaminated soil or can release contaminants or chemical amendments into clean soil

Easy to construct
- available design drawings for construction of future units
- standard engineering equipment for easy fabrication and maintenance

Easy to operate and maintain
- standard engineering equipment and components for easy maintenance
- designed for easy access to controls, gauges, and filters
- manual outlines operations and maintenance

Affordable
- no overhead costs, only budget for shipping and maintenance costs
- no permanent ECRS staff
- no travel costs for researcher, staff, and their instrumentation
- minimal staff needed to operate programmable equipment
- reduced regulatory requirements, only spill containment and appropriate waste treatment

Faster tests
- reduced or eliminated regulatory permitting
- ready to operate within one week of delivery
- can be programmed to operate 24 hours/day, 7 days/week

Appendices

Appendix A

ECRS Unit 1 Engineering Diagrams

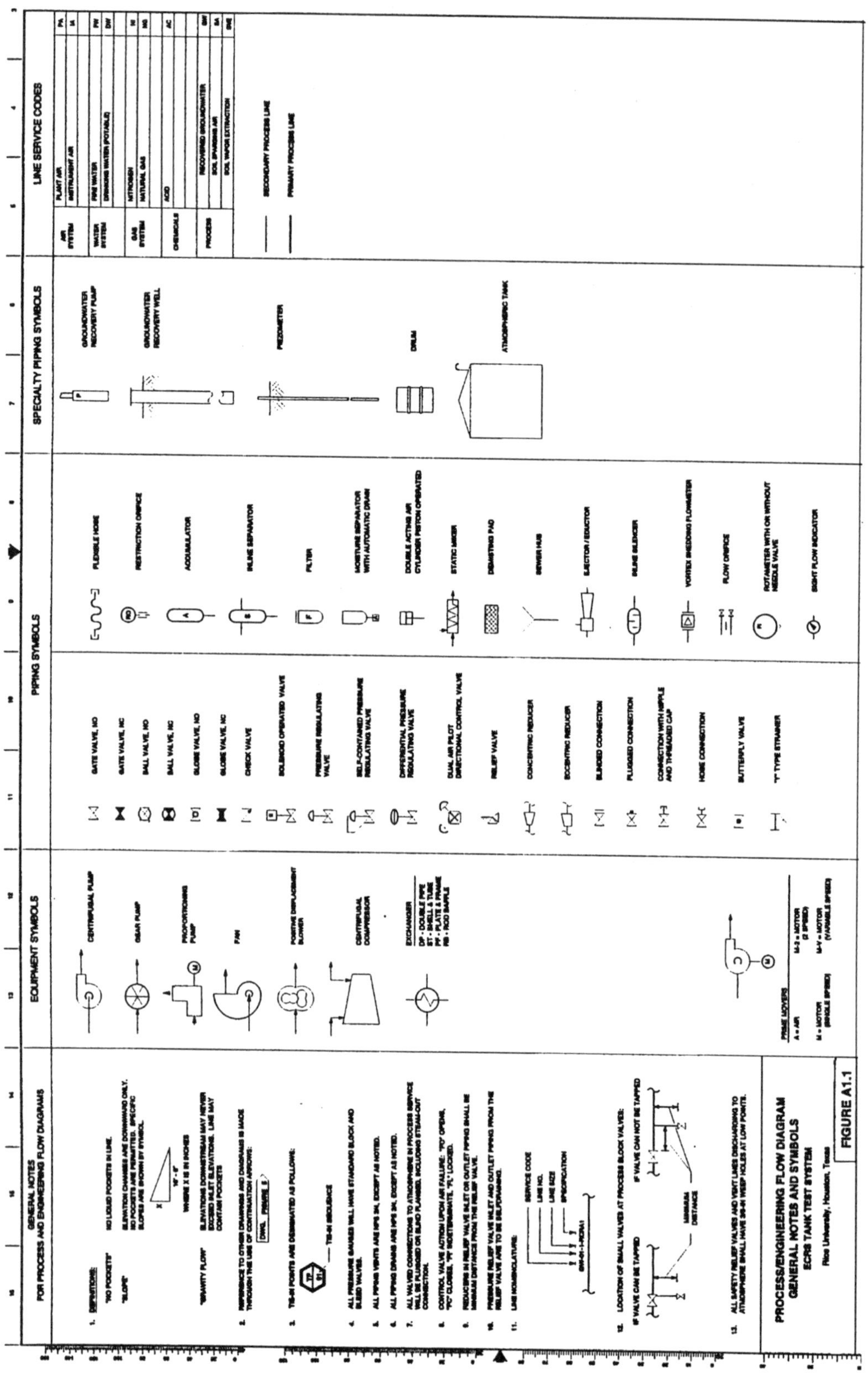

Figure A1.1   Process/engineering flow diagram, general notes and symbols.

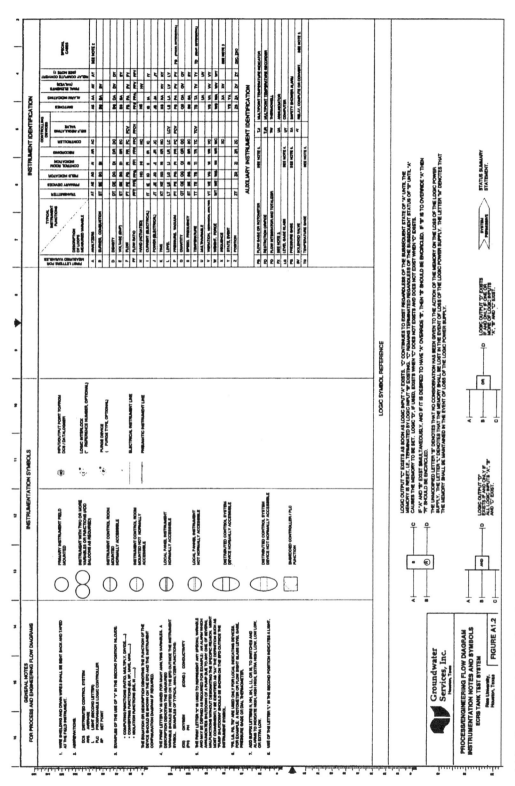

Figure A1.2 Process/engineering flow diagram, instrumentation notes and symbols.

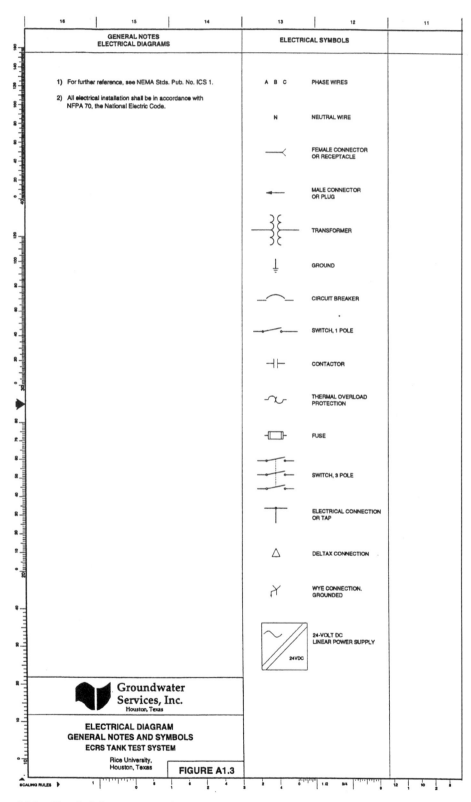

Figure A1.3 Electrical diagram, general notes and symbols.

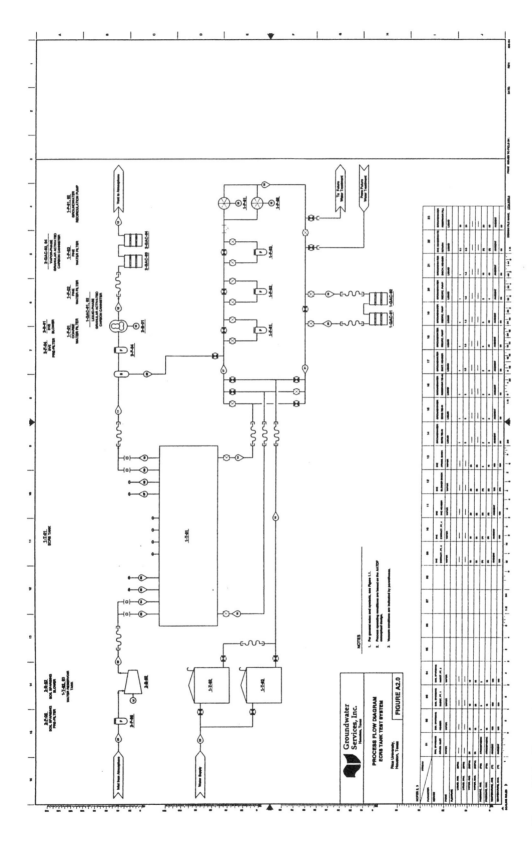

Figure A2.0 Process flow diagram.

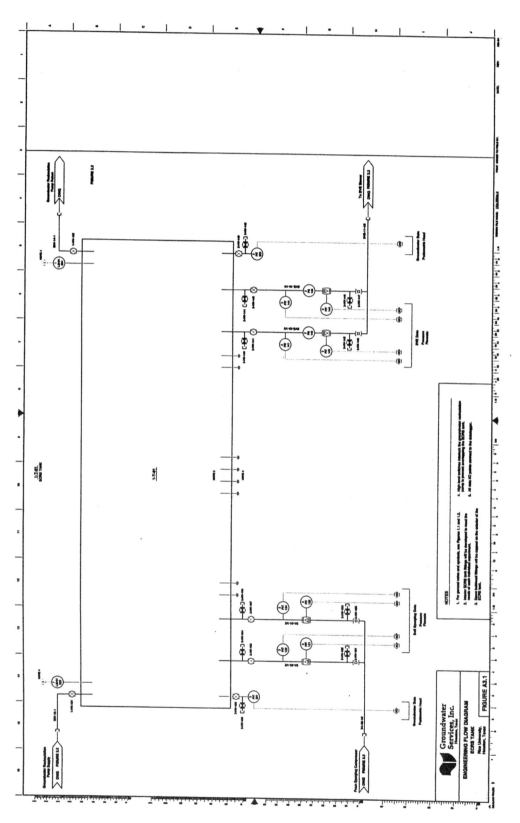

Figure A3.1   Engineering flow diagram, ECRS soil tank.

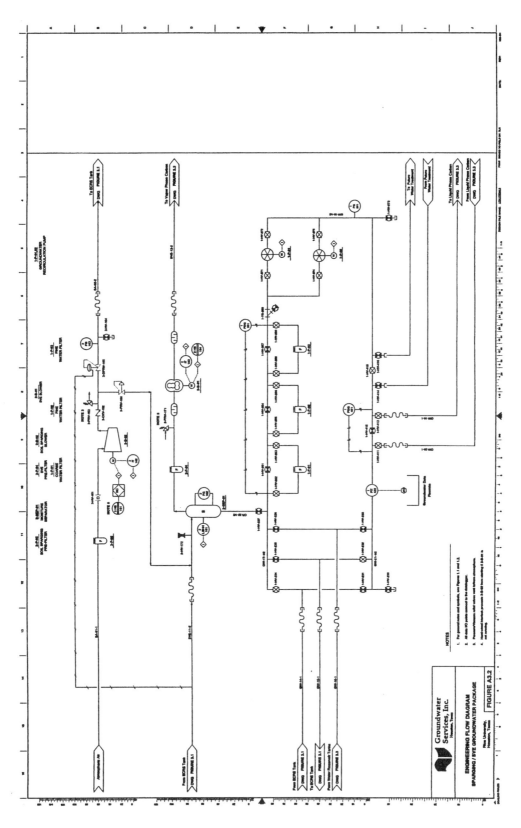

Figure A3.2 Engineering flow diagram, sparging/SVE groundwater package.

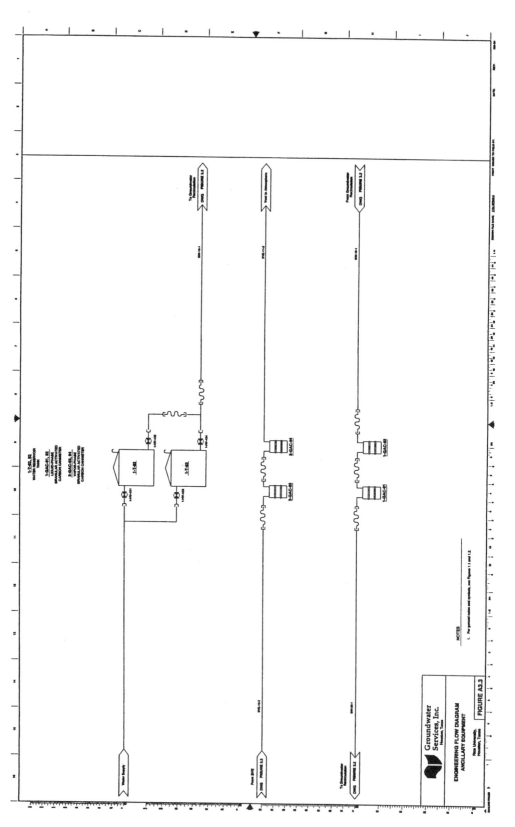

Figure A3.3   Engineering flow diagram, ancillary equipment.

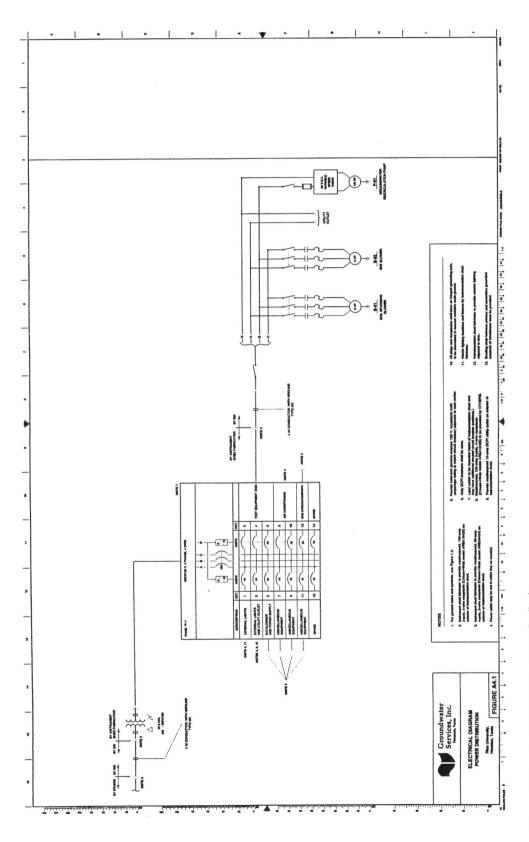

Figure A4.1 Electrical diagram, power distribution.

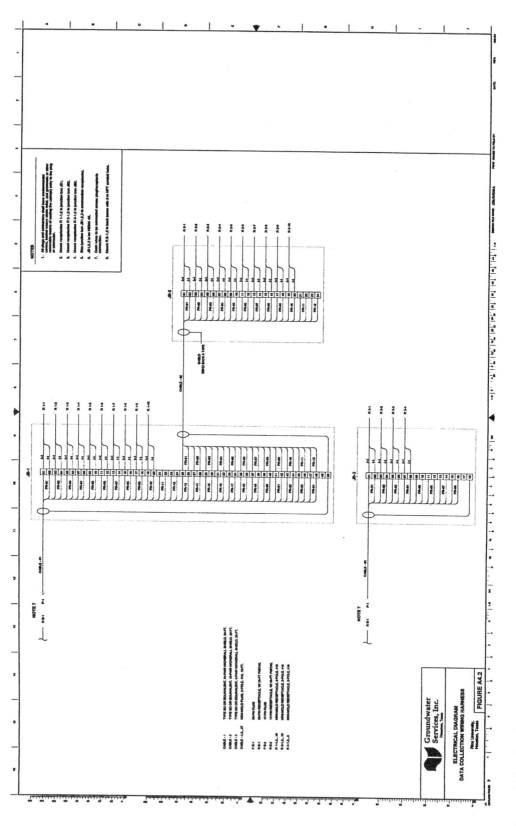

Figure A4.2   Electrical diagram, data collection wiring harness.

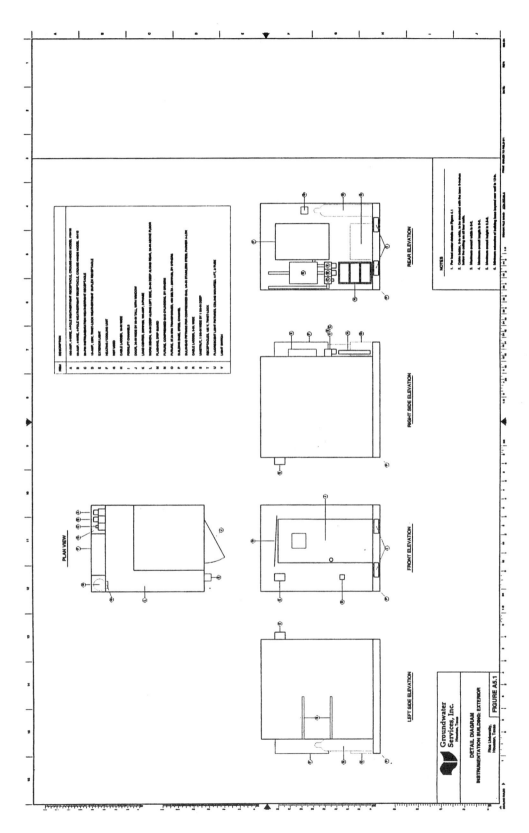

Figure A5.1 Detail diagram, instrumentation building exterior.

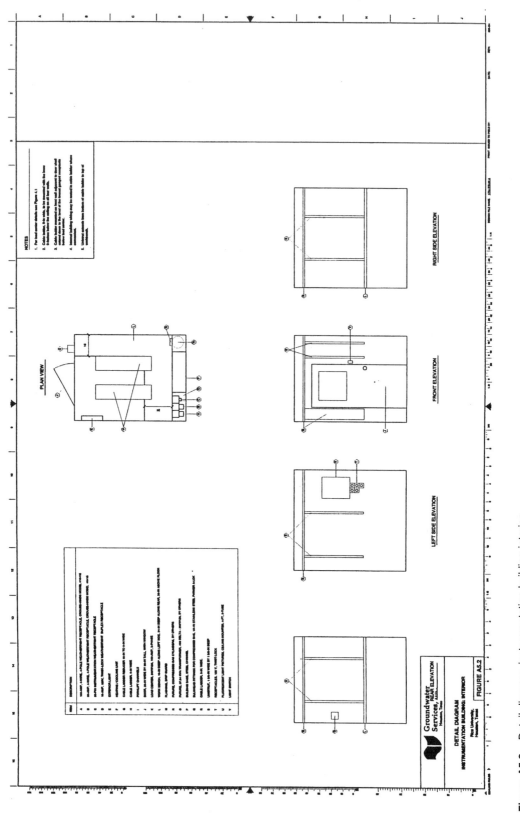

Figure A5.2 Detail diagram, instrumentation building interior.

Appendix B

Example of an ECRS Health and Safety Plan

June 7, 1996

APPENDIX B.1:
EXAMPLE SITE-SPECIFIC
PROJECT HEALTH AND SAFETY PLAN

Experimental Controlled Release System (ECRS)
DOD-AATDF/Rice University, Houston, Texas

Note: In order to prepare a site-specific Health and Safety Plan (HASP), this example form, or the equivalent, should be completed prior to beginning site operations. All field personnel working with ECRS should read, understand, and observe the provisions of the site-specific HASP.

_____

Project Name

_____

Project Location

Job No. _____

Prepared By: _____ Date: _____

Reviewed By: _____ Date: _____

ACKNOWLEDGMENT

I, the undersigned, have been provided with a copy of this Site-Specific Project Health and Safety Plan. I have read the Plan, have attended a project safety orientation session conducted by _____ and have had the opportunity to ask questions about health and safety issues relating to this project. I understand that it is my responsibility to abide by this Plan, and that physical injury, damage and other harm to myself or others could result from my failure to do so.

Signature	Name	Date
Signature	Name	Date
Signature	Name	Date
Signature	Name	Date
Signature	Name	Date

June 7, 1996

APPENDIX B.1:
EXAMPLE SITE-SPECIFIC
PROJECT HEALTH AND SAFETY PLAN

Experimental Controlled Release System (ECRS)
DOD-AATDF/Rice University, Houston, Texas

1.0   PROJECT DESCRIPTION

1.1   Client/Project   _____

_____

1.2   Project Location   _____

1.3   Project Manager   _____

1.4   Project Safety Officer   _____

1.5   Client Safety Contact Person   _____

Address and Phone No.   _____

_____

1.6   Project Activities   _____

_____

_____

2.0   SITE DESCRIPTION AND HAZARD IDENTIFICATION

2.1   Site Description   _____

_____

_____

2.2   Physical and Biological Hazards

_____ Heat Stress   _____ Heavy Equipment Operations

_____ Fire   (Fire Extinguishers Required)   _____ Poisonous Snakes

_____ Other (specify)   (e.g., electric shock) _____

2.2   Potential Chemical Exposure   _____

_____

APPENDIX B.1:
EXAMPLE SITE-SPECIFIC
PROJECT HEALTH AND SAFETY PLAN

Experimental Controlled Release System (ECRS)
DOD-AATDF/Rice University, Houston, Texas

PRINCIPAL CHEMICALS OF CONCERN

CHEMICAL NAME	PEL (ppm/8 hrs)*	STEL (ppm/15 min)*	IDLH (ppm)	REMARKS

Notes:
1. A Material Safety Data Sheet for each chemical of concern should be appended to the HASP.
2. * = Units and duration unless otherwise noted.
   NPV = No published value.

3.0   HEALTH AND SAFETY PROTOCOL

3.1   Site Control

_____   Barricades (Specify) _____

_____   Vapor Control (Specify Method and Application)

_____   Other (Specify) _____

3.2   Training Required (Specify OSHA HAZWOPER or other)

June 7, 1996

APPENDIX B.1:
EXAMPLE SITE-SPECIFIC
PROJECT HEALTH AND SAFETY PLAN

Experimental Controlled Release System (ECRS)
DOD-AATDF/Rice University, Houston, Texas

3.3   Air Quality Monitoring

3.3.1   Air Monitoring Instrument

_____ OVA   _____ OV Badges (Specify Compounds) _____
_____ Draeger Tube (specify compound) _____
_____ Other (Specify) _____

3.3.2   Monitoring Frequency and Zone

_____ Work Zone _____ Site Perimeter ____ Other (specify)
Frequency: ____ Continuously   _____ Hourly
            ____ Noticeable Odor   _____ Other (Describe)

3.4   Personal Protective Equipment Level

3.4.1   EPA Equipment Level Required ____ B   ____ C   ____ D

Note: For all Level "A" tasks and Level "B" tasks involving or use of heavy equipment, formal safety plan documents must be prepared.

3.4.2   Level "D" Equipment Specifications

____ Hard Hat   ____ Goggles   _____ Safety Glasses

____ Surgical Gloves   ____ Cotton Work Gloves

____ Neoprene, Nitrile, or Rubber Gloves   ____ Hearing Protection

____ Steel-Toed Boots (Rubber, PVC, or Nalgene)

____ Tyvek   ____ Poly-Coat Tyvek   ____ Face Shield

____ Other Equipment (Specify)   Life lines _____

Conditions For Upgrade to Level "C"   _____

_____

June 7, 1996

APPENDIX B.1:
EXAMPLE SITE-SPECIFIC
PROJECT HEALTH AND SAFETY PLAN

Experimental Controlled Release System (ECRS)
DOD-AATDF/Rice University, Houston, Texas

3.4.3   Level "C" Equipment Specifications
(Specify respirator use in addition to Level "D" equipment above).

_____ Half-Face Respirator   _____ Full Face Respirator

Specify Cartridge Type _____

_____ Other Equipment (Specify)   _____

Conditions For Upgrade to Level "B"   _____

_____

3.4.3   Level "B" Equipment Specifications
(Specify supplied air use in addition to Level "D" equipment above).

_____ Continuous Flow   _____ Pressure Demand

_____ Other Equipment (Specify)   _____

Conditions For Project Shut-down   _____

3.5   Decontamination Protocol

Personnel _____

_____

_____

PPE Disposal _____

Equipment _____

_____

3.6   Location of Emergency Medical Facilities (Attach Map)

_____

_____

_____

June 7, 1996

APPENDIX B.1:
EXAMPLE SITE-SPECIFIC
PROJECT HEALTH AND SAFETY PLAN

Experimental Controlled Release System (ECRS)
DOD-AATDF/Rice University, Houston, Texas

3.7   Accident Reporting Procedures

_____

_____

3.8   Chemical Spill Response and Reporting Procedures

_____

_____

4.0   UNDERGROUND UTILITIES CLEARANCE

4.1   Underground Utilities Board Contacted:

	Case No.	Date
Texas One Call System (800) 245-4545	_____	_____
Texas Utilities Coordinating Committee (713) 223-4567	_____	_____

4.2   Site Clearance Obtained: _____   _____
                                               (name)   (date)

4.3   Utilities Checked:

_____   Electric   _____

_____   Gas   _____

_____   Telephone   _____

_____   TV Cable   _____

_____   Water/Sewer   _____

June 7, 1996

APPENDIX B.1:
EXAMPLE SITE-SPECIFIC
PROJECT HEALTH AND SAFETY PLAN

Experimental Controlled Release System (ECRS)
DOD-AATDF/Rice University, Houston, Texas

4.4   Petroleum Pipelines (List)

_____   _____

_____   _____

_____   _____

_____   _____

_____   _____

_____   _____

5.0   MISCELLANEOUS
(Special instructions, remarks, etc.   Attach additional sheets as needed).

1)   Standard lockout/tagout out procedures will be followed while working
with electrical equipment.

2)   Only personnel trained in confined space entry will enter the ECRS tank.

3)   Standard line breaking procedures will be followed when breaking any
piping is separated for maintenance or repair work.

_____

_____

_____

_____

_____

_____

June 7, 1996

APPENDIX B.2:
EXAMPLE PROCEDURE FOR
CONFINED SPACE ENTRY

Experimental Controlled Release System (ECRS)
DOD-AATDF/Rice University, Houston, Texas

1.0   PURPOSE

This procedure is established to comply with OSHA Regulations (29 CFR 1910.146 (c)(4) concerning confined space entry. This procedure establishes the minimum requirements for the entrance into a confined space. It shall be used to ensure that potentially hazardous asphyxiating, toxic, flammable/explosive atmospheres, engulfment hazards, stored energy sources, and physical hazards have been isolated or removed prior to entrance into a confined space for construction, service, or maintenance activities.

2.0   SCOPE

This procedure applies to all employees of the Host Facility and their subcontractors involved in construction, service, or maintenance activities which involve entry into the ECRS Tank.

3.0   DEFINITIONS

3.1   Confined Space

A space large enough and so configured that an employee can bodily enter and perform assigned work. In addition, a confined space has limited or restricted means for entry or exit, and is not designed for continuous employee occupancy. A confined space is defined as any work area that has, or may have, one or more characteristics, as follows:

- *Hazardous Atmosphere:* A confined space that contains or has the potential to contain a hazardous atmosphere. This could mean that the oxygen content of the space is inadequate, or that toxic or explosive gases, fumes, or vapors are present.

- *Potential for Engulfment:* A confined space that contains a material that has the potential for engulfing an authorized entrant (e.g. loose soils, grain, water).

June 7, 1996

APPENDIX B.2:
EXAMPLE PROCEDURE FOR
CONFINED SPACE ENTRY

Experimental Controlled Release System (ECRS)
DOD-AATDF/Rice University, Houston, Texas

- *Potential for Asphyxiation:* A confined space that has an internal configuration such that an entrant could be trapped or asphyxiated by inwardly converging walls or by a floor which slopes downward and tapers to a smaller cross-section.

3.2   Entry

The action by which a person passes through an opening into a permit-required confined space. Entry includes ensuing work activities in that space and is considered to have occurred as soon as any part of the entrant's body breaks the plane of an opening into the space.

3.3   Entry Permit

The written or printed document that is provided by the employer to allow and control entry into a permit space.

3.4   Attendant

An individual stationed outside of one or more permit spaces who monitors the authorized entrants and who performs all the attendant's duties assigned in GSI's permit space program.

3.5   Authorized Entrant

An employee who is authorized by the employer to enter a permit space.

3.6   Entry Supervisor

The person responsible for the following: i) determining if acceptable entry conditions are present at the permit space where entry is planned; ii) authorizing entry; iii) overseeing entry operations; and iv) terminating entry as required. The duties of the entry supervisor, attendant and authorized entrant may be passed from one individual to another during the course of the entry operation as long as that individual has been trained and equipped as required for each role to be filled.

June 7, 1996

APPENDIX B.2:
EXAMPLE PROCEDURE FOR
CONFINED SPACE ENTRY

Experimental Controlled Release System (ECRS)
DOD-AATDF/Rice University, Houston, Texas

4.0   RESPONSIBILITIES

4.1   Employer Responsibilities

The employer is responsible for preparation of a confined space procedure in conformance with 29 CFR 1910.146 (c)(4), including proper training of its authorized employees affected by this procedure, and for providing necessary safety equipment.

4.2   Attendant Responsibilities

- Know the hazards that may be faced during entry of the confined space, including how exposure occurs, and the signs, symptoms, and consequences of exposure.
- Be aware of possible behavioral effects of hazard exposure in authorized entrants.
- Continuously maintain an accurate count of authorized entrants in the permit space and ensure that the permit accurately identifies the authorized entrants who are in the permitted space.
- Remain outside the permit space during entry operations until relieved by another attendant.
- Communicate with authorized entrants as necessary to monitor their status and to alert them of the need to evacuate the permit space.
- Monitor activities inside and outside the space to determine if it is safe for authorized entrants to remain in the space and order the authorized entrants to evacuate the permit space immediately under any of the following conditions: if a prohibited condition is detected, if behavioral effects of hazardous exposure are observed in an authorized entrant, if a situation outside the permit space that could endanger the authorized entrants is observed, or if the attendant can not effectively and safely perform all of his or her duties.
- Summon rescue and other emergency services as soon as it is determined that authorized entrants may need assistance to escape from the permit space.
- Take the following action when unauthorized persons approach or enter a permit spaced while entry is underway: warn the unauthorized persons that they must stay away from the permit space, advise the unauthorized persons that they must exit immediately if they have entered the permit space, inform

APPENDIX B.2:
EXAMPLE PROCEDURE FOR
CONFINED SPACE ENTRY

Experimental Controlled Release System (ECRS)
DOD-AATDF/Rice University, Houston, Texas

the authorized entrants and entry supervisor if unauthorized persons have entered the permit space.
- Perform non-entry rescues as specified by this procedure.
- Perform no other duties that might interfere with the primary duty to monitor and protect the authorized entrants.

4.3   Authorized Entrant Responsibilities

- Know the hazards that may be faced during entry of a permit space, including how exposure occurs, as well as the signs, symptoms, and consequences of exposure.
- Properly use equipment as required by this procedure.
- Communicate with the attendant as needed so that the attendant can monitor your status, and alert you of the need to evacuate the permit space.
- Alert the attendant whenever you recognize any warning sign or symptom of exposure to a dangerous situation or a prohibited condition.
- Exit a permit space as quickly as possible whenever the following occurs: an order to evacuate is given by the attendant or the entry supervisor, you recognize any warning sign or symptom of exposure to a dangerous situation, a prohibited condition is observed, or an evacuation alarm is activated.

4.4   Entry Supervisor Responsibilities

- Know the hazards that may be faced during entry of a permit space, including how exposure occurs, and the signs, symptoms, and consequences of exposure.
- Before the entry permit can be signed, and prior to entry, the following must be verified: all tests specified by the permit have been conducted, all procedures and equipment specified by the permit are in place.
- Terminate the entry and cancel the permit as required by this procedure.
- Verify that rescue services are available and that a procedure exists for summoning them when necessary.
- Remove unauthorized individuals who enter or attempt to enter the permit space during entry operations.
- Determine that entry proceedings remain consistent with the terms of the entry permit and that acceptable entry conditions are maintained whenever responsibility for a permit space entry operation is transferred, and at

APPENDIX B.2:
EXAMPLE PROCEDURE FOR
CONFINED SPACE ENTRY

Experimental Controlled Release System (ECRS)
DOD-AATDF/Rice University, Houston, Texas

intervals dictated by the hazards and operations performed within the permit space.

5.0   POLICY

Prior to entrance into the ECRS Tank at the Host Facility, the following procedure will be followed.

5.1   Procedure

The entry supervisor will complete the confined space pre-entry checklist prior to any authorized entry into the ECRS Tank (sample attached). Entry will not proceed until all of the applicable action items on the checklist have been completed. Once all of the action items on the checklist have been completed, the entry supervisor will completely fill-out the Confined Space Entry Permit (sample attached), and prominently display the permit at the ECRS Tank.

5.2   Work Plan

Dirt will be placed into the ECRS Tank by a forklift with a bucket attachment. Under no circumstances will authorized entrants be inside of the tank while dirt is being placed. After the forklift has backed away from the tank, the LEL meter will be lowered into the tank with a tag-line and then walked the length of the tank to determine if an oxygen deficient atmosphere is present (less than 19.5 % oxygen). If an oxygen deficient atmosphere is present, entry into the tank is forbidden. If adequate levels of oxygen are present, authorized entrants may enter the tank by a ladder from the scaffolding. The ladder must extend a minimum of 3 feet above the top of the tank. All equipment and tools will be lowered into and lifted out of the tank with tag-lines.

Authorized entrants shall wear a full face respirator equipped with dust filter cartridges, hard hat, hearing protection, and full body harness with retrieval line attached at the center D-ring of the entrant's back (near shoulder level). Only two authorized entrants may be in the tank at any time. At least one attendant must be present when an authorized entrant is in the tank. The attendant will attach the retrieval lines of the authorized entrants to the forklift. The work

June 7, 1996

APPENDIX B.2:
EXAMPLE PROCEDURE FOR
CONFINED SPACE ENTRY

Experimental Controlled Release System (ECRS)
DOD-AATDF/Rice University, Houston, Texas

space of the permit space will be continuously monitored for the percent oxygen. If an oxygen deficient atmosphere is ever detected, authorized entrants will immediately leave the ECRS Tank.

This confined space procedure will be followed until the working level in the ECRS Tank is less than four feet in depth.

5.3   Rescue Plan

The on-site rescue team will consist of those employees who are outside of the permit space. Retrieval lines attached to the forklift will be utilized to remove unconscious entrants from the tank and onto the scaffold decking. Entry of the rescuer into the permit space to help remove an injured entrant is discouraged. Each member of the rescue service must be trained in basic first-aid and in cardiopulmonary resuscitation (CPR). At least one member of the on-site rescue service must hold current certification in both first-aid and CPR.

June 7, 1996

APPENDIX B.3:
EXAMPLE PROCEDURE FOR
LINE BREAKING

Experimental Controlled Release System (ECRS)
DOD-AATDF/Rice University, Houston, Texas

1.0   PURPOSE

This procedure establishes the minimum requirements for line breaking for maintenance and repair operations. It will be used to prevent injuries, spills, fires, and explosions during line breaking operations.

2.0   SCOPE

This procedure applies to all employees of the Host Facility and its subcontractors involved with the separation of pipelines, transfer lines and any other permanent, fixed, flexible or temporary piping. It does not apply to hoses used in the temporary connection of a tank truck, vacuum truck, or pressure washer.

This procedure represents the minimum requirements for line breaking. In most cases the operating facility where the work is taking place will have it's own line breaking procedure which takes precedence over this standard operating procedure (SOP).

3.0   RESPONSIBILITIES

3.1   Project Manager

- Ensure that only properly trained employees perform line breaking activities and that Line Breaking Permits are properly completed before work begins.

3.2   Site Safety Officer

- Review and approve Line Breaking Permits.
- Inspect activities to ensure that they are performed safely and in compliance with this procedure.
- Provide training in safe work practices for line breaking.
- Ensure that subcontractors have equivalent training before performing line breaking work.

June 7, 1996

APPENDIX B.3:
EXAMPLE PROCEDURE FOR
LINE BREAKING

Experimental Controlled Release System (ECRS)
DOD-AATDF/Rice University, Houston, Texas

3.3   All Other Employees

- Complete required training before participating in line breaking work.
- Perform line breaking activities safely and in compliance with this procedure.

4.0   EQUIPMENT REQUIREMENTS

The following equipment is required before line breaking work can begin: lockout devices and tags (see Lockout/Tagout SOP), buckets, pails, spill absorbent and containers, shovel, fire extinguisher, first aid kit, emergency shower and eyewash.

5.0   TRAINING

All employees must be trained in safe work practices for line breaking before participating in this work activity. Training will be presented by Host Facilities Safety Officer or designee and consist of classroom lecture and demonstration exercises. Records of line breaking training will be kept in the project files.

6.0   OPERATING PROCEDURE

6.1   Pre-Mobilization

- Identify the materials that are or have been present in each line to be broken. Obtain Material Data Safety Sheet(s) or equivalent information on each material from Project Manager.

- Contact Project Manager to arrange for equipment and process lockouts during line breaking activities. Notify other site personnel involved in advance of any utility or system shutdown, so that critical machinery and processes are not damaged during line breaking.

- Determine whether the lines to be opened have been drained and purged. If so, determine when and how this was done.

June 7, 1996

APPENDIX B.3:
EXAMPLE PROCEDURE FOR
LINE BREAKING

Experimental Controlled Release System (ECRS)
DOD-AATDF/Rice University, Houston, Texas

- For gases, determine where excess gas will be vented. Never release toxic or reactive gases into the work environment. Releases to the outside air must not violate local, state or federal air pollution controls. The Project Manager will determine which standards apply to the specific project.

- Assemble lockout and spill control equipment that will be required for the specific job.

6.2   Line Breaking Procedure for Liquids

- Drain all lines back into their respective storage reservoirs or tanks, if possible.

- Shut off and lock out all connections and valves for the lines to be broken. Shut off and lockout all power and electrical supplies for the lines in question to prevent refilling or repressurizing the lines when they are open (see Lockout/Tagout SOP). *Warning:* Double check for cross-connections and backup devices that could refill the lines.

- Check line pressure valves, if available, to determine whether each line is still under pressure.

- Don appropriate Personal Protective Equipment (PPE) as defined in the site-specific Health and Safety Plan.

- Place a bucket or other appropriate container under the drain valve for each line. Be sure that the bucket or container is chemically compatible with the liquid to be drained.

- Remove all unnecessary personnel from the immediate work area. Take a position standing to the side of the valve to avoid being sprayed by the material being drained from the line.

- Drain excess liquid from appropriate valves on all liquid lines. Locate low points in the lines where liquids may have accumulated and drain those lines too.

June 7, 1996

APPENDIX B.3:
EXAMPLE PROCEDURE FOR
LINE BREAKING

Experimental Controlled Release System (ECRS)
DOD-AATDF/Rice University, Houston, Texas

- Flush the lines with water or another purging solution that is chemically compatible with the liquid in the lines. Inert gas, such as nitrogen, can also be used, but gases are not as effective for purging liquid lines.

- Drain and flush the lines a minimum of three times, collecting all flushing solution in appropriate containers for disposal.

- Verify that there is no pressure remaining on the lines to be broken.

- Complete the Line Breaking Permit and obtain approval for the Site Safety Officer to proceed.

- Lines that have contained potentially flammable or reactive materials should be filled with an inert gas before cutting to prevent fires and explosions. Use only non-sparking tools and cold-cutting methods to cut these lines.

- Place a bucket or catch basin under the point to be broken in the line to catch any residual liquid.

- Stand to the side of lines being broken to avoid contact with any residual material in the lines.

6.3   Line Breaking Procedure for Gases

- Drain all lines back into their respective storage reservoirs or tanks, if possible.

- Shut off and lock out all connections and valves for the lines to be broken. Shut off and lockout all power and electrical supplies for the lines in question to prevent refilling or repressurizing the lines when they are open (see Lockout/Tagout SOP). *Warning:* Double check for cross-connections and backup devices that could refill the lines.

- Check line pressure valves, if available, to determine whether each line is still under pressure.

June 7, 1996

APPENDIX B.3:
EXAMPLE PROCEDURE FOR
LINE BREAKING

Experimental Controlled Release System (ECRS)
DOD-AATDF/Rice University, Houston, Texas

- Don appropriate Personal Protective Equipment (PPE) as defined in the site-specific Health and Safety Plan.

- Remove all unnecessary personnel from the immediate work area. Take a position standing to the side of the valve to avoid being sprayed by the gas if the valve fails.

- Vent any remaining pressure on lines into an appropriate receiver. Recheck pressure valves to see that lines have reached ambient air pressure.

- Flush the lines at least three time with an inert gas, such as nitrogen, to remove residual gas from low points in the lines.

- Complete the Line Breaking Permit and obtain approval from the Site Safety Officer to proceed.

- Lines that have contained potentially flammable or reactive materials should be filled with an inert gas before cutting to prevent fires and explosions. Use only non-sparking tools and cold-cutting methods to cut these lines.

- Stand to the side of the line being broken to avoid contact with any residual material in the line.

June 7, 1996

APPENDIX B.4:
EXAMPLE PROCEDURE FOR
ELECTRICAL/HAZARDOUS ENERGY LOCKOUT

Experimental Controlled Release System (ECRS)
DOD-AATDF/Rice University, Houston, Texas

1.0   PURPOSE AND SCOPE

1.1   Purpose

This procedure is established to comply with OSHA Regulations (29 CFR §1910.147) concerning control of hazardous energy. The procedure establishes the minimum requirements for the lockout or tagout of energy isolating devices during maintenance and repair operations. It shall be used to ensure that machines and electrical equipment are isolated from all sources of potentially hazardous energy, and are locked out or tagged out before employees perform service or maintenance activities where the unexpected energizing, startup, or release of stored energy could cause injury.

1.2   Scope

This procedure applies to all employees of the Host Facility and their subcontractors involved in the installation, service, and maintenance of powered machinery and equipment. Such equipment currently in use in the ECRS is limited to electric-powered pumps, compressor, and blower, with associated motor control equipment.

Accordingly, this procedure is limited to the lock out of electrical equipment and does not cover the lockout of hydraulic-powered or pneumatic-powered equipment, or lock out of radioactive energy sources. All employees and subcontractors are specifically instructed not to enter any vessel or activate any valves or other devices relating to hydraulic, pneumatic, radioactive, or process chemical systems under any circumstance. Should scope of work be expanded in the future such that access to these types of systems becomes necessary, this procedure will be revised and expanded as appropriate to comply with OSHA standards before such work is performed.

June 7, 1996

APPENDIX B.4:
EXAMPLE PROCEDURE FOR
ELECTRICAL/HAZARDOUS ENERGY LOCKOUT

Experimental Controlled Release System (ECRS)
DOD-AATDF/Rice University, Houston, Texas

2.0   DEFINITIONS

2.1   Affected employee

An employee whose job requires him/her to operate or use a machine or electrical equipment on which servicing or maintenance is being performed under lockout or tagout, or whose job requires him/her to work in an area in which servicing is being performed.

2.2   Authorized employee

A person who implements a lockout/tagout system procedure on machines or equipment to perform maintenance or service on that machine or equipment.

2.3   Capable of being locked out

An energy isolating device is considered capable of being locked out if it is designed with a hasp or other attachment or integral part to which, or through which a lock can be affixed, or if it has a locking mechanism built into it.

2.4   Energized

Connected to an energy source or containing residual or stored energy.

2.5   Energy isolating device

A mechanical device that physically prevents the transmission or release of energy, including, but not limited to the following: a manually operated electrical circuit breaker, a disconnect switch, or a manually operated switch by which the conductors of a circuit can be disconnected from all ungrounded supply conductors.

2.6   Lockout

Placement of a lockout device on an energy isolation device ensuring that the energy isolating device and the equipment being controlled can not be operated until the lockout device is removed.

June 7, 1996

APPENDIX B.4:
EXAMPLE PROCEDURE FOR
ELECTRICAL/HAZARDOUS ENERGY LOCKOUT

Experimental Controlled Release System (ECRS)
DOD-AATDF/Rice University, Houston, Texas

2.7   Lockout Device

A device that utilizes a positive means, such as a keyed lock, to hold an energy isolating device in the safe position and prevent the energizing of a machine or equipment. For the purpose of this procedure, a keyed padlock, color-coded OSHA yellow, identified as a lockout device, and labeled with the name of the owning authorized employee will be used.

2.8   Machinery or Equipment

As used in this procedure, machinery or equipment refers to any mechanical or electrical device containing or utilizing potentially hazardous energy, including pumps, motors, etc.

2.9   Tie Wrap

A non-reusable self-locking device, such as a one-piece nylon cable tie, which is attachable by hand and non-releasable with a minimum unlocking strength of 50 pounds.

2.10   Tagout

The use of lockout tags and tie wraps on energy sources which can not be locked out. A tagout is not as secure as a lockout and may only be used when a lockout is not feasible.

3.0   RESPONSIBILITIES

3.1   Employer Responsibilities

The employer is responsible for preparation of a lockout procedure in conformance with the 29 CFR § 1910.147, including proper training of its authorized and affected employees in that procedure, and for providing an adequate supply of locks and other necessary lockout devices.

June 7, 1996

APPENDIX B.4:
EXAMPLE PROCEDURE FOR
ELECTRICAL/HAZARDOUS ENERGY LOCKOUT

Experimental Controlled Release System (ECRS)
DOD-AATDF/Rice University, Houston, Texas

3.2   Authorized Employee Responsibilities

The authorized employee is responsible for safe and proper implementation of this lockout procedure (i.e., for preventing accidental operation or energizing of machinery or equipment by proper use of lockout devices).

3.3   Individual/Affected Employee Responsibilities

Persons performing inspection or repair work are responsible for their own safety and protection by having satisfied themselves that the authorized employee has correctly locked out the equipment to be repaired. The authorized employee will assist the individual in reviewing the lockout plan, verifying the lockout, and testing start switches after lockout. It is mandatory that the maintenance and or operations person try the start switch or switches to verify that the proper equipment has been locked out and that there are no additional energy sources.

4.0   POLICY AND PROCEDURES

4.1   Lockout/Tagout Procedure

Prior to any maintenance or service to be performed on equipment covered by this procedure (i.e., electric pumps, compressor, blower, and associated motor control equipment), the authorized employee will notify all affected employees of the lockout. The equipment will then be shut down and locked out by the authorized employee.

The lock may be placed on the main power switch located outside the circuit breaker box, on the lockout/tagout service shutoff for the SVE and sparging system located behind the control panel or the applicable switch has been switched to the off position, in the case of the water pumps.

Each lockout device will have a proper lockout tag stating the reason for the lockout, the name of the authorized employee performing the lockout, the date of the lockout and, if the equipment is to be out of service beyond that date, the expected time period of the lockout.

June 7, 1996

APPENDIX B.4:
EXAMPLE PROCEDURE FOR
ELECTRICAL/HAZARDOUS ENERGY LOCKOUT

Experimental Controlled Release System (ECRS)
DOD-AATDF/Rice University, Houston, Texas

Acceptable lockout devices include color-coded keyed padlocks with a label identifying it as a lockout device and bearing the name of the authorized employee. Other devices such as multi-holed hasps (if two or more employees are to work on the locked out equipment) and circuit breaker lockouts may be used in conjunction with the padlocks. Each authorized employee will be issued an adequate number of keyed locks and will retain sole possession of the keys. All lockout equipment will be substantial enough to prevent removal without the use of excessive force or unusual techniques such as bolt cutters or other metal cutting tools. In cases where placement of a lock is not possible, a tagout label on a tie wrap may be substituted.

4.2 Verification of Deactivation/De-energizing

After all sources of energy have been locked/tagged out, the authorized employee will verify that the equipment has been de-energized by an appropriate method. For equipment on which a switch is located downline of the lockout point, the switch will be tried prior to proceeding with the work. A volt meter will be used to verify the absence of potentially hazardous energy from electrical connections prior to disconnection.

4.3 Equipment Startup

At the completion of the job, the authorized employee is responsible for verification that the job is complete and the need for the lockout is over. At this point, the job site will be inspected to verify that restart of the equipment will not endanger personnel. All lockout devices will be removed by the authorized employee(s). The equipment may then be restarted.

4.4 Application of General Procedure

Electrical equipment currently in use is limited to pumps, compressor, blower and related motor control equipment. This ECRS system does not involve maintenance or repairs to systems utilizing chemical, thermal, pneumatic, hydraulic, or radioactive energy. In addition, equipment operated by meeting the following criteria, specified in 29 CFR § 1910.147 (c)(4), permitting application

APPENDIX B.4:
EXAMPLE PROCEDURE FOR
ELECTRICAL/HAZARDOUS ENERGY LOCKOUT

Experimental Controlled Release System (ECRS)
DOD-AATDF/Rice University, Houston, Texas

of a single general lockout procedure and providing an exemption from development of a specific procedure for each piece of equipment:

- The machine or equipment has no potential for residual or stored energy or re-accumulation of stored potentially hazardous energy after shut down;

- The machine or equipment has a single energy source which can readily be identified and isolated;

- The isolation and locking out of that energy source will completely de-energize and deactivate the machine or equipment;

- The machine or equipment is isolated from that energy source and locked out during servicing or maintenance;

- A single lockout device will achieve a lockout condition;

- The lockout device is under the exclusive control of the authorized employee performing the servicing or maintenance;

- The servicing or maintenance does not create other hazards for other employees;

- The Host Facility, in utilizing this exception, has had no accidents involving the unexpected activation or re-energizing of the machine or equipment during servicing or maintenance.

Accordingly, the lockout procedure described above will be followed for maintenance and service of all pumps, compressor, blower, and related equipment.

4.5   Inadvertent/Abandoned Locks

In the event that a lockout is inadvertently left on equipment or is abandoned, the lock may be removed by the Host Facilities safety officer or his/her designee. An attempt will be made to immediately notify the individual of the lock removal, and a written record made of the removal, including the location of the

APPENDIX B.4:
EXAMPLE PROCEDURE FOR
ELECTRICAL/HAZARDOUS ENERGY LOCKOUT

Experimental Controlled Release System (ECRS)
DOD-AATDF/Rice University, Houston, Texas

lockout, the individual who performed the lockout, and the individual(s) authorizing and performing the removal. The record will be maintained by the Host Facilities company safety officer.

4.6   Training

All affected and authorized employees will be thoroughly trained in this procedure, and refresher training will be conducted annually at a minimum. Any employee and /or subcontractors performing work for the Host Facility under a lockout will also be trained in the procedure, as needed. Additional training will be performed as necessary (e.g., in the event of deficiencies in implementation). Records of training will be maintained by the Host Facilities company safety officer.

4.7   Verification of Compliance

The lockout procedure will be reviewed annually, at a minimum, to verify that the procedure is being followed and to assess the need for revision of the procedure. In addition, inspections will be conducted annually to ensure the procedure is being properly implemented.

Appendix C

ECRS Unit 1 Systems Tests and Research by Equilon (Shell Development Co.)

Note: Figures from Jonathan T. Miller and Ira J. Dortch, April, 1998. Results from the Experimental Controlled Release System (ECRS Unit 1), Westhollow Technology Center, Shell Development Co.

Experimental Controlled Release System (ECRS)

Side view

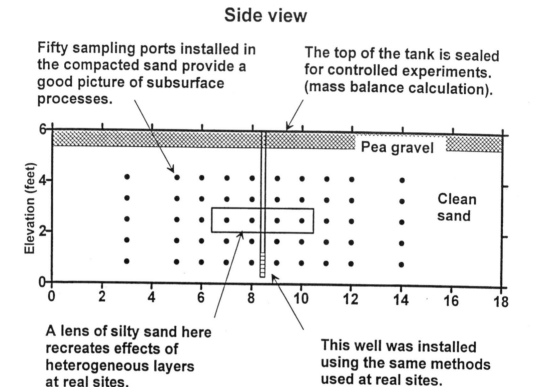

Fifty sampling ports installed in the compacted sand provide a good picture of subsurface processes.

The top of the tank is sealed for controlled experiments. (mass balance calculation).

Pea gravel

Clean sand

A lens of silty sand here recreates effects of heterogeneous layers at real sites.

This well was installed using the same methods used at real sites.

Figure C1.0   Soil tank cross-section view with location of sampling ports.

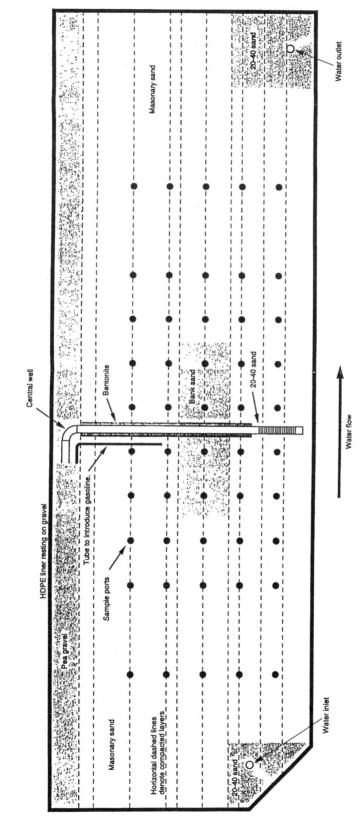

Figure C2.0   Soil tank internal packing.

Grain Size Distribution

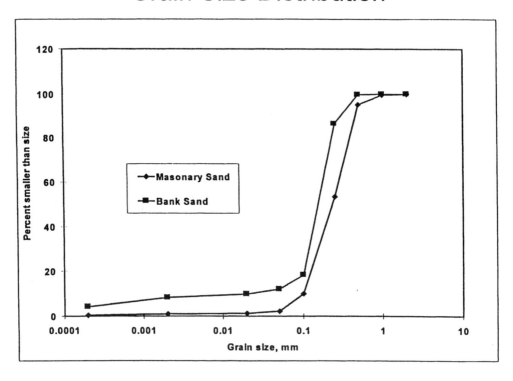

Figure C3.0   Grain size distribution for soil pack.

ECRS central well construction

As built. Drawing not to scale.

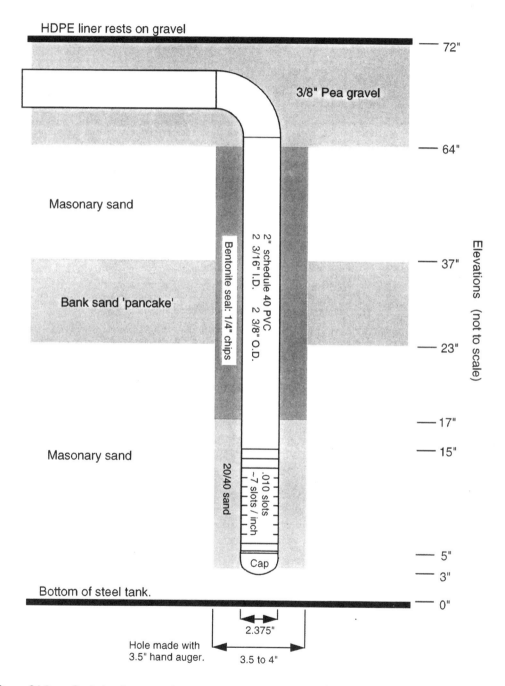

Figure C4.0 Central well construction.

Air Flow Test Configuration

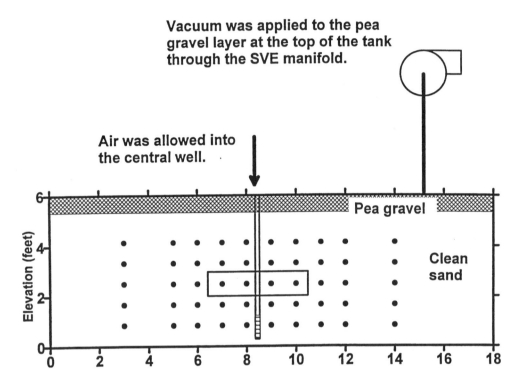

Vacuum was applied to the pea gravel layer at the top of the tank through the SVE manifold.

Air was allowed into the central well.

Pea gravel

Clean sand

Elevation (feet)

Figure C5.0   Air flow test configuration.

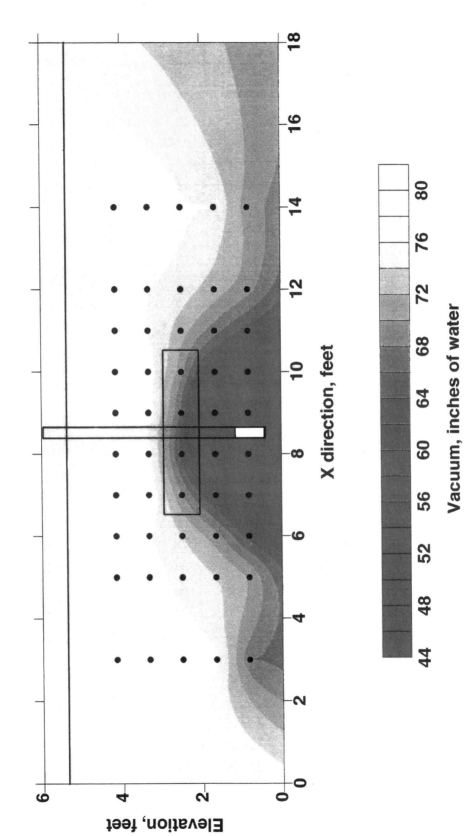

Figure C6.0   Air flow test vacuum results.

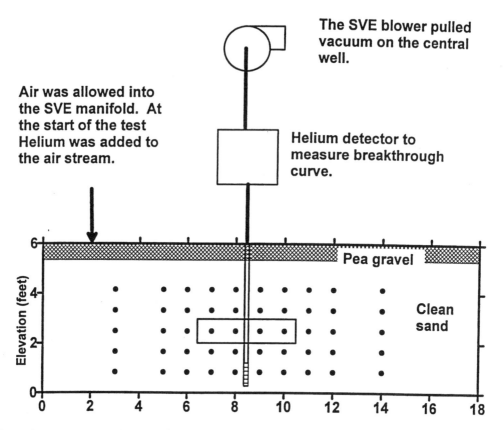

Figure C7.0   Helium leak test configuration.

Helium Leak Test Results

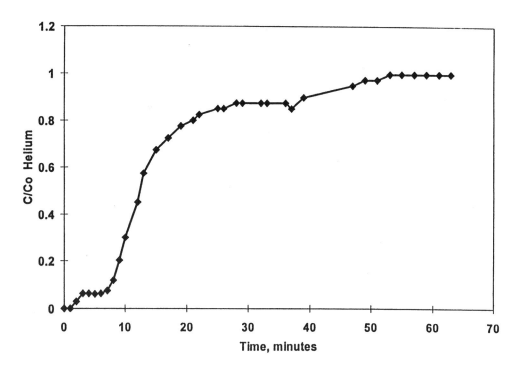

Figure C8.0   Helium leak test results.

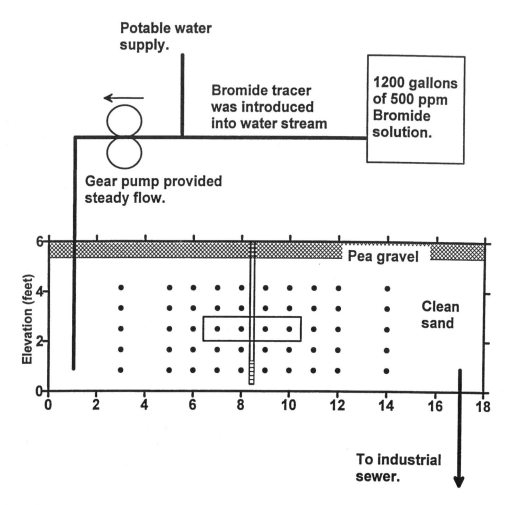

Figure C9.0   Water tracer test configuration.

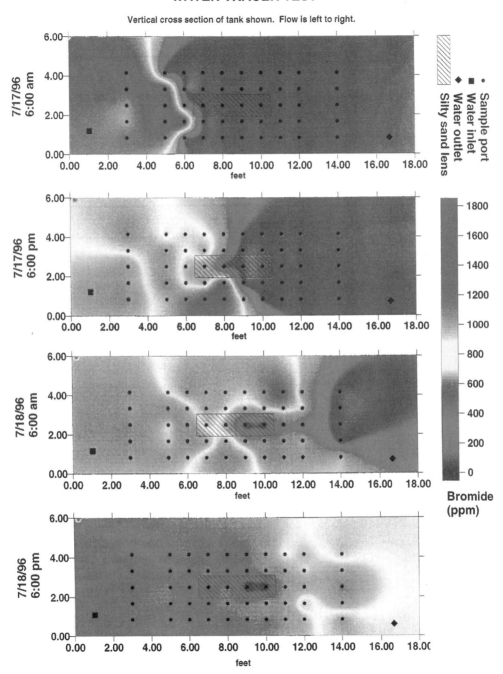

Figure C10.0   Water tracer test results.

Natural Attenuation Experiment Configuration

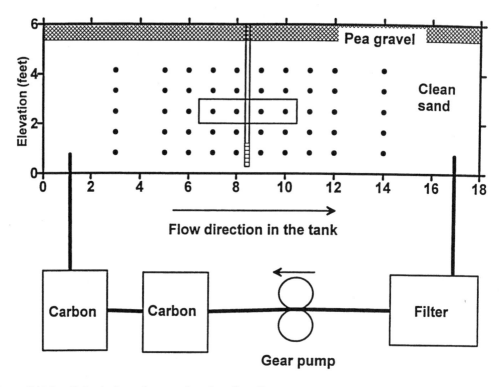

Figure C11.0   Natural attenuation experiment configuration.

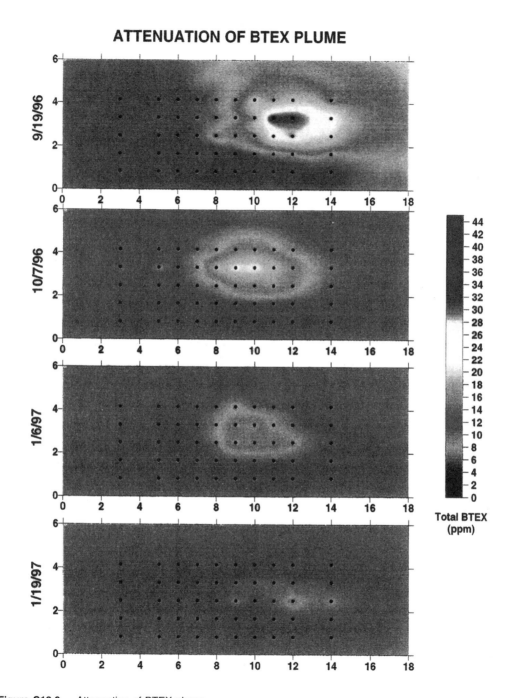

Figure C12.0   Attenuation of BTEX plume.

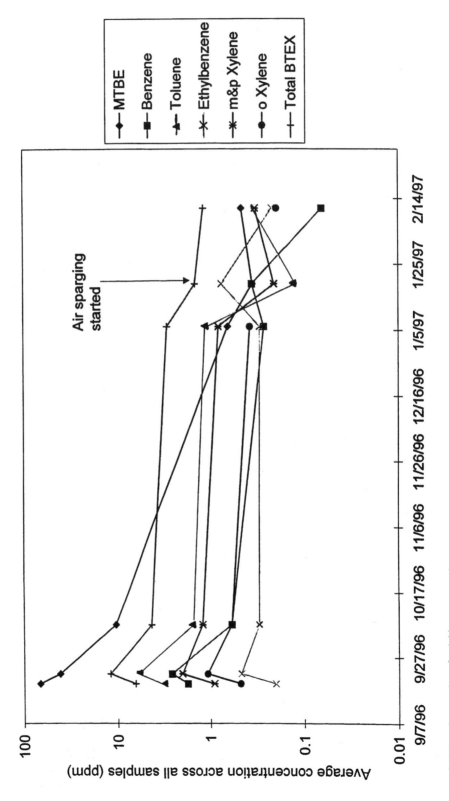

Figure C13.0   Attenuation of soluble components.

Air Sparging Experiment Configuration

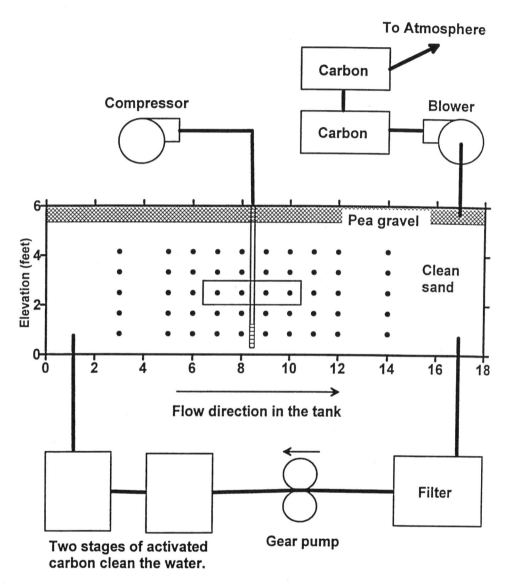

Figure C14.0   Air sparging experiment configuration.

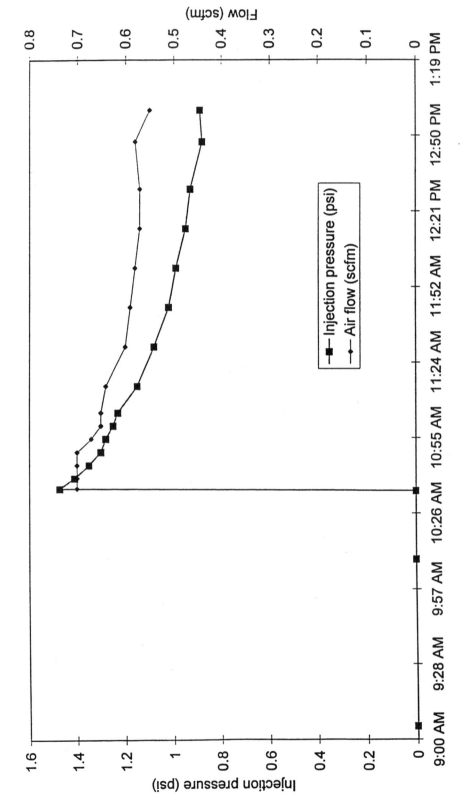

Figure C15.0 Air sparging pressure and flow.

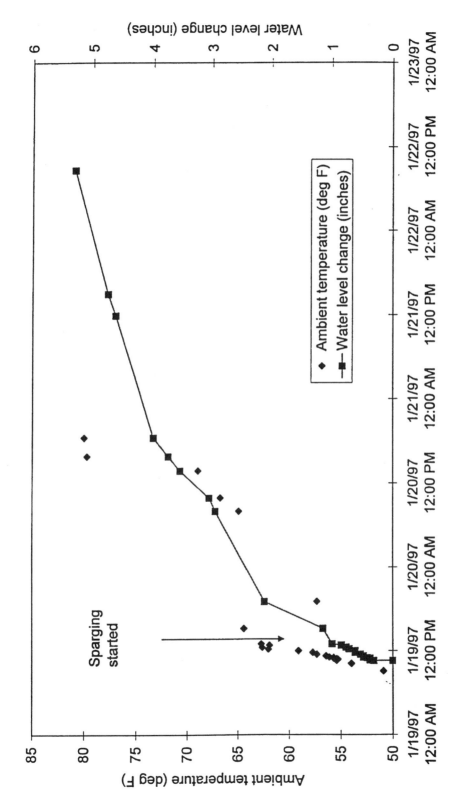

Figure C16.0   Water level and temperature during air sparging.

Changes in Dissolved Oxygen Due To Air Sparging

Dissolved Oxygen before sparging

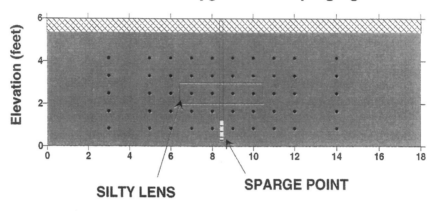

SILTY LENS   SPARGE POINT

Dissolved Oxygen 20 hours after sparging started.

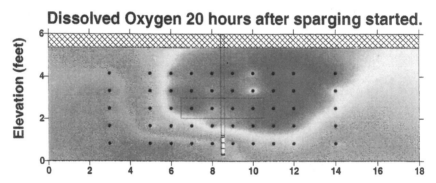

Dissolved Oxygen after 24 days of steady sparging.

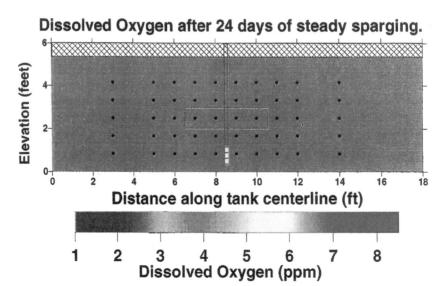

Distance along tank centerline (ft)

1   2   3   4   5   6   7   8

Dissolved Oxygen (ppm)

Figure C17.0   Changes in dissolved oxygen due to air sparging.

ORM Experiment Configuration

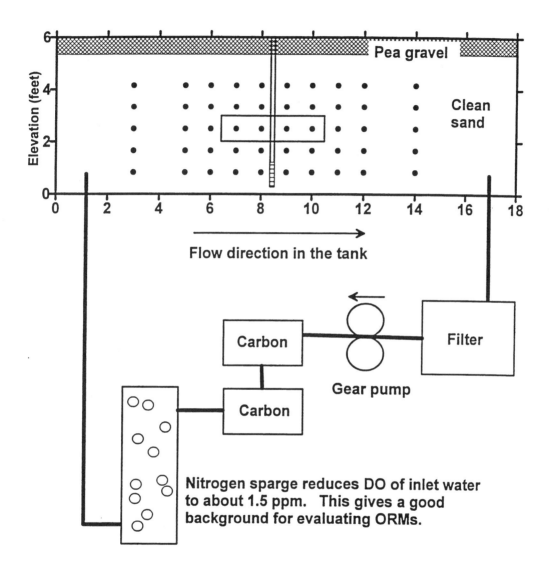

Figure C18.0   ORM experiment configuration.

Appendix D

ECRS Unit 2 Bid Package and Engineering Specifications

GSI Job No. G-1874
September 13, 1996

TABLE OF CONTENTS:
ENGINEERING SPECIFICATIONS

Experimental Controlled Release System (ECRS) II
DOD-AATDF/Rice University, Houston, Texas

1.0 INSTRUCTIONS TO BIDDERS

Tables

Figures

2.0 SPECIFICATIONS: GENERAL

Specifications

Figures

GSI Job No. G-1874
September 13, 1996

TABLE OF CONTENTS:
ENGINEERING SPECIFICATIONS

Experimental Controlled Release System (ECRS) II
DOD-AATDF/Rice University, Houston, Texas

3.0   SPECIFICATIONS: PROCESS EQUIPMENT SKID

Specifications

Figures

GSI Job No. G-1874
September 13, 1996

TABLE OF CONTENTS:
ENGINEERING SPECIFICATIONS

Experimental Controlled Release System (ECRS) II
DOD-AATDF/Rice University, Houston, Texas

4.0 SPECIFICATIONS: ECRS TANK

Specifications

Figures

5.0 SPECIFICATIONS: ANCILLARY EQUIPMENT

Specifications

Figures

GSI Job No. G-1874
September 13, 1996

1.0   INSTRUCTIONS TO BIDDERS

1.1   Project Overview

The Department of Defense Advanced Applied Technology Demonstration Facility for Environmental Technology (DOD-AATDF) began in May 1993 when the DOD awarded a grant to a consortium of six university-based environmental research centers. The goal of the AATDF project is to sponsor studies and activities that are designed to provide the missing link between technology development and application. The AATDF is administered by the U.S. Army Corps of Engineers Waterways Experiment Station (WES) and is located within the Brown School of Engineering at Rice University.

As part of the AATDF program, ECRS II has been designed to facilitate quantitative assessment of the effectiveness of various remediation techniques involving saturated or unsaturated subsurface conditions on a pilot scale of operations. The ECRS tank and connections are configured to facilitate various schemes for sampling, injection, and extraction of water and/or air. As shown on Figure 1.1, primary system components comprise 1) a process equipment skid containing an air compressor, water pumps, controls, and instrumentation; 2) a 27 cubic yard rectangular tank modified with nozzles, piping, and valves; and 3) a ancillary tanks, pumps, and hose connections. Design basis information and major equipment for the system are summarized on Tables 1.1 and 1.2, respectively.

A previous DOD/AATDF project, (i.e., ECRS I completed in June 1996) has some attributes in common with the ECRS II project described in this engineering specification (See Figure 1.2). Please note that these photographs are provided for reference only; the ECRS II project shall be constructed in full accordance with the requirements detailed in this engineering specifications document.

1.2   General Requirements

1.2.1   Scope of Work

With the exception of those work items specifically excluded per Section 1.2.2 below, the Contractor shall provide all labor, equipment, materials, and expertise and complete all related tasks required for completion of ECRS II in full accordance with the standards specified in this specifications document.

GSI Job No. G-1874
September 13, 1996

1.2.2   Work Excluded

As detailed on the attached drawings and Bid Quotation Form (see Table 1.3), Rice will purchase and supply to the Contractor specific equipment and instrumentation required for project completion. Rice will also provide certain items of equipment for storage by the Contractor (see Section 1.5.4 below). No other equipment, materials, or labor will be provided by Rice.

The completed system as installed may include treatment for air or water extracted from the ECRS tank. As described in Section 5.0, the Contractor shall provide connections to such treatment facilities, however, the Contractor will not be responsible for the treatment units.

1.2.3   Governing Codes

The work to be performed shall be accomplished in accordance with referenced codes; local, state and federal regulations; and procedures covered in the specifications.

1.2.4   Project Coordination

Engineering oversight during construction will be provided by a designated representative of Rice University. The Contractor will accommodate scheduled inspections during construction and deliveries to the installation site with the Rice Representative throughout the duration of the work program. A minimum of two inspections shall be conducted: the first at the midpoint of construction and the second prior to shipment. Other inspections may be scheduled on an as-needed basis.

1.3   Schedule

A pre-bid meeting will be held on Thursday, September 26, 1996, at 2:00 p.m. at the offices of Groundwater Services, Inc., to review of the engineering specifications. Bid quotations must be submitted on the enclosed Bid Quotation Form (see Table 1.3) by no later than 4:00 pm on October 4, 1996. Project authorization by Rice University is scheduled to be completed by October 11, 1996. All work items included in this specification must be completed by January 8, 1997, and delivered to the final delivery destination by January 15, 1997.

GSI Job No. G-1874
September 13, 1996

1.4   Acceptance of Work

1.4.1   Required Testing

Prior to shipment, the Contractor shall demonstrate to the satisfaction of the Rice Representative that the process equipment skid, ECRS tank, and ancillary equipment have been completed in accordance with applicable specifications. For the ECRS tank and ancillary equipment, this demonstration will consist of a visual inspection by the Rice Representative in conjunction with a review of relevant testing documentation (see Section 1.5.2 below).   In the case of the process equipment skid, the Contractor shall operate the equipment, instrumentation, and controls on the process equipment skid in order to verify achievement of design basis conditions (see Table 1.1). This test of the process equipment skid shall be conducted in the presence of the Rice Representative. The Contractor shall also demonstrate that the all input and output signal cables to and from instrumentation, equipment, and the data recorder and all power cables are properly terminated per relevant specifications.

1.4.2   Required Documentation

The Contractor shall provide documentation that relevant components on the process equipment skid, the ECRS tank, and ancillary equipment have been pressure tested in accordance with applicable requirements (see Specifications 2.1, 2.2, and 2.3). All original documentation received from equipment vendors and suppliers, including but not limited to certifications, operation manuals, data sheets, shop drawings, etc., shall be transferred to the Rice Representative at the time of final inspection.

1.4.3   Storage

The Contractor shall provide space for temporary storage of two items to be provided by Rice, as follows: 1) a portable instrumentation building, and 2) an unassembled portable shelter. The space shall be a secure indoor area that is locked when unoccupied by Contractor personnel. The footprint dimensions of the instrumentation building will not exceed 8 ft wide and 10 ft long; the height will not exceed that allowable for transport by a standard flatbed truck without a special permit, and the weight will not exceed 2,000 lb. The dimensions of the unassembled portable shelter will not exceed 6 ft wide, by 8 ft long, by 8 ft high.

1.4.4   Delivery

The Contractor shall provide for delivery of the completed work items described in this engineering specification (i.e., the process equipment skid, the ECRS tank,

GSI Job No. G-1874
September 13, 1996

and the ancillary equipment), the instrumentation building, and the unassembled shelter. All items to be delivered shall be configured and packed to fit on one flatbed truck for transport. Shipment dates shall be coordinated with the Rice Representative. The Contractor will be responsible for all equipment until inspected and accepted by the Rice Representative at the final delivery destination, at which time unloading and installation of these items will be completed by others.

Final delivery of the completed process equipment skid, ECRS tank, ancillary equipment, instrumentation building, and unassembled shelter will be made to one of two locations: 1) Arizona State University in Tempe, Arizona; or 2) the University of Texas in Austin, Texas. The bidder is requested to provide delivery costs to both locations (see Table 1.3, Bid Quotation Form). The Contractor will be advised of the delivery location as soon as possible during the construction phase of the project.

SECTION 1.0 TABLES

Table 1.1   Design Basis Information
Table 1.2   List of Major Equipment
Table 1.3   Bid Quotation Form

GSI Job No. G-1874
Issued: 9/13/96
Page 1 of 1

TABLE 1.1
DESIGN BASIS INFORMATION

Experimental Controlled Release System (ECRS) II
Rice University, Houston, Texas

SYSTEM	DESCRIPTION	PARAMETER	RANGE
Water Recirculation	Water is circulated through the ECRS tank by applying gradient control using pressure regulators. Water in excess of that required to maintain the defined hydraulic gradient is returned to a surge tank prior to filtration, any additional treatment (i.e., carbon adsorption to remove organic compounds), and recirculation to the upgradient overflow weir.	Groundwater Flowrate Groundwater Elevation (above bottom of tank) Hydraulic Gradient	0 - 10 gpm 1 - 6 ft 0 - 0.4 ft/ft
Air Sparging	Filtered compressed air is regulated and injected into the ECRS tank. The injection pressure is referenced to the SVE vacuum and is designed to maintain a constant differential pressure across the tank.	Air Flowrate Injection Pressure (relative to applied SVE vacuum)	0 - 30 scfm 0 - 5 psi difference
Soil Vapor Extraction (SVE)	A venturi vacuum pump driven by compressed air provides SVE capability for the system. The motive air pressure is regulated to apply a constant vacuum at the top of the ECRS tank. Entrained solids and water are removed in a centrifugal separator upstream of the vacuum pump.	Air flowrate SVE vacuum	0 - 30 scfm 0 - 15 in Hg vacuum
Chemical Injection	A mixing tank and metering pump are provided to facilitate the controlled addition of water soluble tracers and chemicals to the groundwater supply stream or to specific point(s) in the process.	Chemical injection flowrate	0 - 7 gph

GSI Job No. G-1874
Issued: 9/13/96
Page 1 of 1

TABLE 1.2
MAJOR EQUIPMENT

Experimental Controlled Release System (ECRS) II
DOD-AATDF/Rice University, Houston, Texas

ITEM NO.	DESCRIPTION	REFERENCE
C-1	Air Compressor	Specification 3.1
F-1	Coalescing Filter	Specification 3.2
F-2	Carbon Filter	Specification 3.2
F-3, 4	Groundwater Filters	Specification 3.6
P-1	Groundwater Recirculation Pump	Specification 3.5
P-2	Utility Pump	Specification 5.1
P-3	Chemical Injection Pump	Specification 5.1
SP-1	Venturi Pump	Specification 3.4
SP-2	Venturi Pump	Specification 3.4
T-1	ECRS Tank	Figures 4.2
T-2	Water Reservoir	Specification 5.2, Figure 5.2
T-3	Chemical Mixing Tank	Specification 5.2, Figure 5.2
V-1	Air Receiver	Specification 3.2
V-2	Moisture Separator	Figure 3.2

Note:

1. Locations of above referenced equipment are shown on Engineering Flow Diagrams for process equipment skid, ECRS tank, and ancillary equipment (Figures 3.1, 4.1, and 5.1, respectively). Referenced specifications and manufacturers' data are provided in Sections 3 - 5 of this document.

TABLE 1.3
BID QUOTATION FORM

Experimental Controlled Release System (ECRS) II
DOD-AATDF/Rice University, Houston, Texas

BIDDER IDENTIFICATION:

Company Name: _____

Company Address: _____

                   _____

Contact Name: _____

Telephone: _____ Fax: _____

BID CONDITIONS

Work tasks and cost figures presented in this quotation are based upon careful review of the specifications and guidelines detailed in "Engineering Specifications, Experimental Controlled Release System (ECRS) II, Rice University, Houston, Texas," issued September 13, 1996.

BID INFORMATION

The following additional items are enclosed with this bid submittal:

1. Contractor Schedule
2. Description of any recommended alternates to specifications (optional).
3. Exceptions to scope of work (optional)

Failure to provide all specified information will disqualify this bid.

COSTS

The following costs are lump sum quotations based on the requirements of the Engineering Specifications document. Lump sum costs specified in this quotation were made after making any and all inquiries, measurements, and investigations necessary to complete the work in accordance with the specifications of the Engineering Specifications document.

TABLE 1.3
BID QUOTATION FORM

Experimental Controlled Release System (ECRS) II
DOD-AATDF/Rice University, Houston, Texas

WORK DESCRIPTION	TOTAL COST
Item 1: *Process Equipment Skid*	
1A Labor: Lump sum cost for fabrication of process equipment skid	$_____
1B Materials, Instrumentation: Lump sum cost for purchase of major instrumentation for process equipment skid: • Data recorder (UIR-1) • Pressure transmitters (PIT-1, 2, 3, 4, 5, 6) • Flow transmitters (FIT-3) • Differential pressure transmitters (FIT-1, 2; LC-1) • Electrical components located on and in control panel CP-1.	$_____
1C Materials, Equipment: Lump sum cost for purchase of those items of instrumentation not listed in item 1B and all other associated process equipment skid items, including pumps, compressors, filters, tanks, piping, valves, and gauges. Utility pump (P-02) will be purchased by Rice University and provided to the contractor.	$_____
Item 2: *ECRS Tank*	
Lump sum cost for purchase and modification of ECRS tank (T-1).	$_____
Item 3: *Ancillary Components*	
Lump sum cost for purchase of water reservoir (T-2), mixing tank (T-3), and all associated hoses, valves, and fittings.	$_____
Item 4: *Shipping*	
4A Lump sum cost for delivery of Items 1A, 1B, and 1C; Item 2; Item 3; the instrumentation building; and the unassembled shelter to Tempe, Arizona.	$_____
4B Lump sum cost for delivery of Items 1A, 1B, and 1C; Item 2; Item 3; the instrumentation building; and the unassembled shelter to Austin, Texas.	$_____
Tempe Delivery Total (items 1A, 1B, 1C, 2, 3, and 4A):	$
Austin Delivery Total (items 1A, 1B, 1C, 2, 3, and 4B):	$

TABLE 1.3
BID QUOTATION FORM

Experimental Controlled Release System (ECRS) II
DOD-AATDF/Rice University, Houston, Texas

CONTRACTOR OFFER

The above cost figures are based upon careful review of the above-referenced bid documents and represent our firm quotation for this work, including all labor, equipment, materials, taxes, fees, and any other expenses related to full and timely completion of the specified tasks.

Offer valid until _____ .

	(Company Name)
By:	
Title:	
Date:	

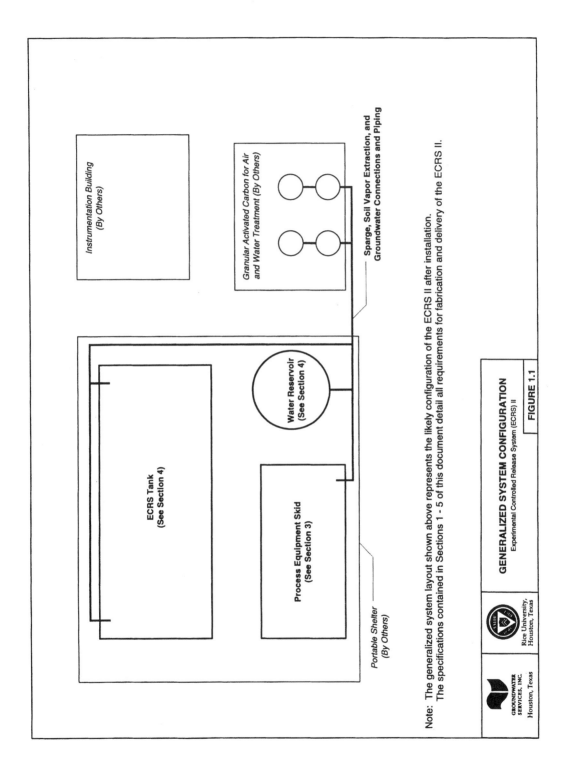

Note: The generalized system layout shown above represents the likely configuration of the ECRS II after installation. The specifications contained in Sections 1 - 5 of this document detail all requirements for fabrication and delivery of the ECRS II.

GENERALIZED SYSTEM CONFIGURATION
Experimental Controlled Release System (ECRS) II

FIGURE 1.1

GROUNDWATER
SERVICES, INC.
Houston, Texas

Rice University,
Houston, Texas

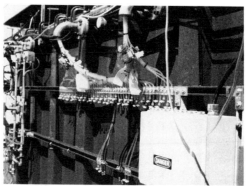

ECRS Unit 1 tank set up at Shell WTC, Houston, Texas

Instrumentation Building

Process Equipment Skid

Water Reservoir Tanks ——

GROUNDWATER SERVICES, INC. Houston, Texas

Rice University
Houston, Texas

PREVIOUS PROJECT EXAMPLE: ECRSI

Experimental
Controlled Release
System (ECRS) II

FIGURE 1.2

GSI Job No. G-1874
September 13, 1996

2.0   GENERAL SPECIFICATIONS

General notes and symbols used for the design and construction drawings in this package are defined on Figures 2.1, 2.2, and 2.3. Piping details are shown on Figure 2.4. A process flow diagram for the system and a general system layout are provided on Figures 2.5 and 2.6, respectively. General piping requirements for stainless steel, flexible hose, and compressed air are provided on piping Specifications 2.1, 2.2, and 2.3, respectively.

GSI Job No. G-1874
September 13, 1996

ENGINEERING SPECIFICATIONS

Experimental Controlled Release System (ECRS) II
DOD-AATDF/Rice University, Houston, Texas

SECTION 2.0 SPECIFICATIONS

Specification 2.1   Piping Specification: AIR1
Specification 2.2   Piping Specification: UTIL3
Specification 2.3   Piping Specification: UTIL4

PIPING AND CONNECTIONS

SIZE	PIPE	FLANGES	FITTINGS	SIZE
1/8 in				1/8 in
1/4 in	Legris Incorporated Rochester, NY 14624 Semi-Rigid Nylon Tubing		Legris Incorporated Rochester, NY 14624 LF3000 Push-In Fittings	1/4 in
3/8 in				3/8 in
1/2 in				1/2 in
3/4 in				3/4 in
1 in				1 in
1-1/2 in				1-1/2 in
2 in				2 in
3 in				3 in
4 in				4 in
5 in				5 in
6 in				6 in

VALVES

SIZE	MANUFACTURER	GLOBE	BALL	CHECK	SPECIAL FEATURES	SIZE
		MODEL NO.				
1/8 in						1/8 in
1/4 in						1/4 in
3/8 in						3/8 in
1/2 in						1/2 in
3/4 in						3/4 in
1 in						1 in
1-1/2 in						1-1/2 in
2 in						2 in
3 in						3 in
4 in						4 in
5 in						5 in
6 in						6 in

SERVICE DESCRIPTION

Fluid: compressed air
Temperature: 0 - 100°F
Pressure: 0 - 80 psi
               0 - 30 psi vacuum

PIPING DETAILS

Bolts: Not applicable
Nuts: Not applicable
Finish/Coating: Not required
Welding: Not applicable

GROUNDWATER SERVICES, INC.
Houston, Texas

Rice University,
Houston, Texas

**PIPING SPECIFICATION AIR 1:
AIR INSTRUMENTATION TUBING**

Experimental Controlled Release
System (ECRS) II

SPECIFICATION 2.1

DESIGN FILE NAME: 1874.AIR1   PRINT ISSUED TO FIELD BY: DATE: REV: K38-D4

NOTES

1. For general notes and symbols, see Figure 2.1.

2. All piping, valves, hoses, and fittings must be assembled in accordance with manufacturers' specifications.

3. Piping and components from alternate manufacturers may be used, provided equivalent or superior performance is obtained.

4. Apply Dow Corning DC-111 silicon compound if necessary to obtain air and vacuum tight seal.

PIPING AND CONNECTIONS

SIZE	PIPE	FLANGES	FITTINGS	SIZE
1/8 in				1/8 in
1/4 in	Type 304 Stainless Steel Tubing		Swagelok Co. Solon, OH 44139 316 Stainless Steel Tube Fittings	1/4 in
3/8 in				3/8 in
1/2 in				1/2 in
3/4 in				3/4 in
1 in	Type 304, Schedule 40S Stainless Steel Seamless Pipe	ANSI Class 150 RFSO, Type 304 Stainless Steel		1 in
1-1/2 in				1-1/2 in
2 in				2 in
3 in				3 in
4 in				4 in
5 in				5 in
6 in				6 in

VALVES

SIZE	MANUFACTURER	MODEL NO.			SPECIAL FEATURES	SIZE
		GLOBE	BALL	CHECK		
1/8 in	Whitey Co. Highland Heights, OH 44143	20 and 26 Series	40 Series			1/8 in
1/4 in						1/4 in
3/8 in	NUPRO Company Willoughby, OH 44094				"TF" Series Tee Type Removable Filter (140 μm wire mesh)	3/8 in
1/2 in						1/2 in
3/4 in						3/4 in
1 in	Neles-Jamesbury, Inc. Worcester, MA 01615		ANSI Class 150, 316 SS, Full Port, Socket Weld Series 4000			1 in
1-1/2 in						1-1/2 in
2 in	Aloyco	ANSI Class 150, 316 SS, Socket Weld				2 in
3 in						3 in
4 in						4 in
5 in						5 in
6 in						6 in

SERVICE DESCRIPTION

Fluid: groundwater, recovered soil vapor
Temperature: 0 - 200°F
Pressure: 0 - 200 psi

PIPING DETAILS

Bolts: Stud bolt ASTM A193, grade B-7
Nuts: ASTM A194, grade 2H
Finish/Coating: Teflon®
Welding: Complete in accordance with ASME B31.3 Section 328 and BPV Code, Section IX.

NOTES

1. For general notes and symbols, see Figure 2.1.

2. During assembly, threads of swaged ferrules shall be sealed with with TFE tape or liquid.

3. All piping, valves, and fittings must be assembled in accordance with manufacturers' specifications.

4. Piping and components from alternate manufacturers may be used, provided equivalent or superior performance is obtained.

GROUNDWATER SERVICES, INC. Houston, Texas
Rice University, Houston, Texas

PIPING SPECIFICATION UTIL3: STAINLESS STEEL TUBING AND PIPING
Experimental Controlled Release System (ECRS) II
SPECIFICATION 2.2

DESIGN FILE NAME: 1874 UTL3   PRINT ISSUED TO FIELD BY: DATE: REV: K96-D4

PIPING AND CONNECTIONS

SIZE	PIPE	FLANGES	FITTINGS	SIZE
1/8 in				1/8 in
1/4 in				1/4 in
3/8 in				3/8 in
1/2 in	Transluscent Polyethlyene Tubing		Dixon Valve & Coupling Co. Chestertown, MD 21620	1/2 in
3/4 in			Type 316 Stainless Steel Shank Type Fittings: Male NPT x Hose Shank	3/4 in
1 in	Dayco Products, Inc. Dayton, OH 45401		Dixon Valve & Coupling Co. Chestertown, MD 21620	1 in
1-1/2 in	POLYCHEM Hose Series 7276	ANSI Class 150 RFSO, Type 304 Stainless Steel	Type 304SS Cam & Groove Couplers: Female Coupler x Hose Shank to Male Adapter x Class 150 ANSI Flange	1-1/2 in
2 in				2 in
3 in				3 in
4 in				4 in
5 in				5 in
6 in				6 in

VALVES

SIZE	MANUFACTURER	GLOBE	BALL	CHECK	SPECIAL FEATURES	SIZE
1/8 in						1/8 in
1/4 in						1/4 in
3/8 in						3/8 in
1/2 in						1/2 in
3/4 in						3/4 in
1 in			ANSI Class 150, 316 Stainless Steel, Full Port, Socket Weld, Series 4000			1 in
1-1/2 in	Neles-Jamesbury, Inc. Worcester, MA 01615					1-1/2 in
2 in						2 in
3 in						3 in
4 in						4 in
5 in						5 in
6 in						6 in

SERVICE DESCRIPTION

Fluid: Groundwater, Recovered Soil Vapor
Temperature: 0 - 200°F
Pressure: 0 - 200 Psi

PIPING DETAILS

Bolts: Stud Bolt ASTM A193, Grade B-7
Nuts: ASTM A194, Grade 2H
Finish/Coating: Teflon®
Welding: Complete In Accordance With ASME B31.3 Section 328 And BPV Code, Section IX.

NOTES

1. For general notes and symbols, see Figure 2.1.

2. All piping, valves, hoses, and fittings must be assembled in accordance with manufacturers' specifications.

3. Piping and components from alternate manufacturers may be used, provided equivalent or superior performance is obtained.

GROUNDWATER SERVICES, INC. Houston, Texas
Rice University, Houston, Texas

PIPING SPECIFICATION UTIL 4: FLEXIBLE HOSE AND TUBING

Experimental Controlled Release System (ECRS) II

SPECIFICATION 2.3

DESIGN FILE NAME: 1674 UTIL4   PRINT ISSUED TO FIELD BY: DATE: REV: IGS-04

GSI Job No. G-1874
September 13, 1996

ENGINEERING SPECIFICATIONS

Experimental Controlled Release System (ECRS) II
DOD-AATDF/Rice University, Houston, Texas

SECTION 2.0 FIGURES

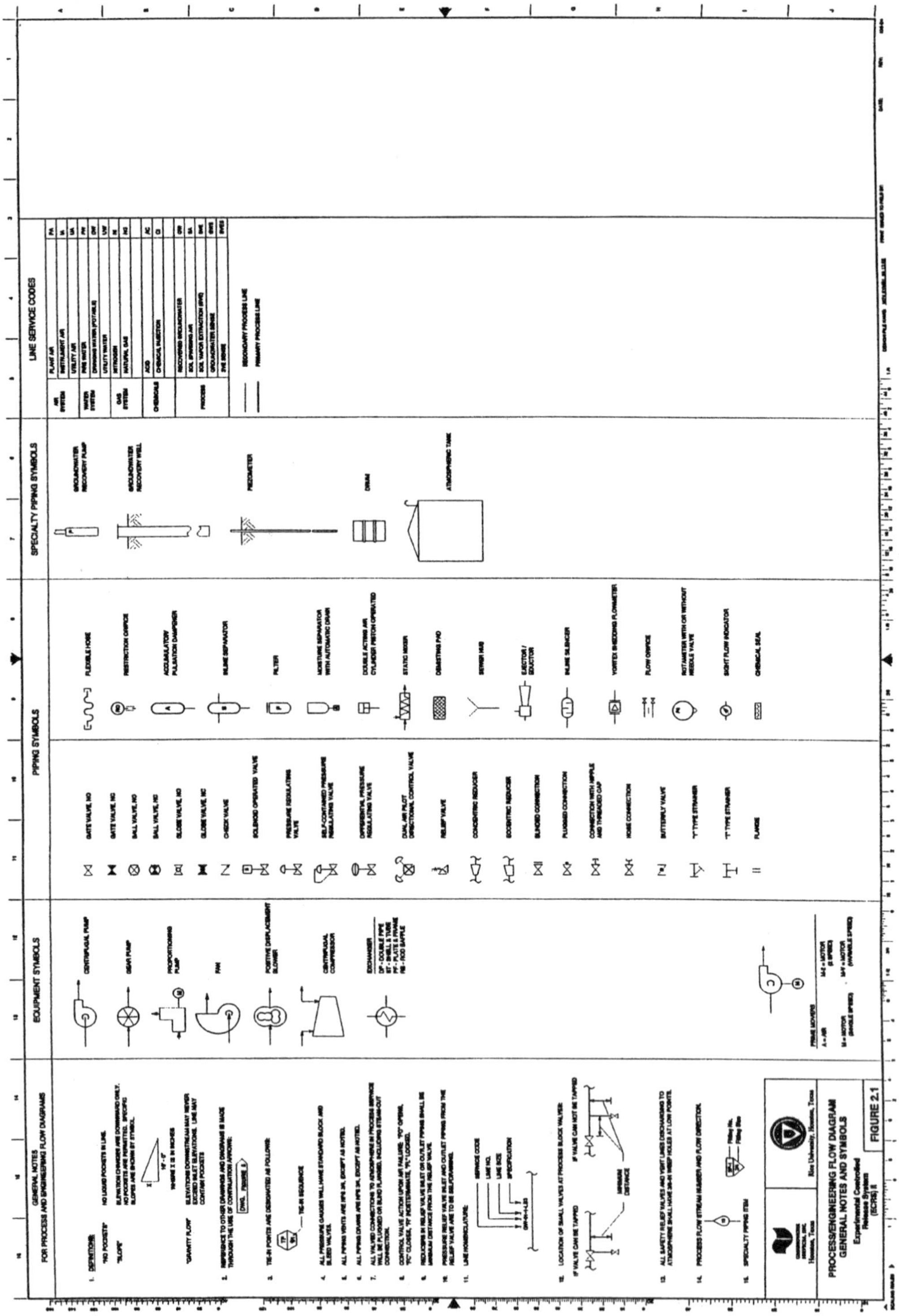
GENERAL NOTES FOR PROCESS AND ENGINEERING FLOW DIAGRAMS
EQUIPMENT SYMBOLS
PIPING SYMBOLS
SPECIALTY PIPING SYMBOLS
LINE SERVICE CODES
PROCESS/ENGINEERING FLOW DIAGRAM
GENERAL NOTES AND SYMBOLS
FIGURE 2.1

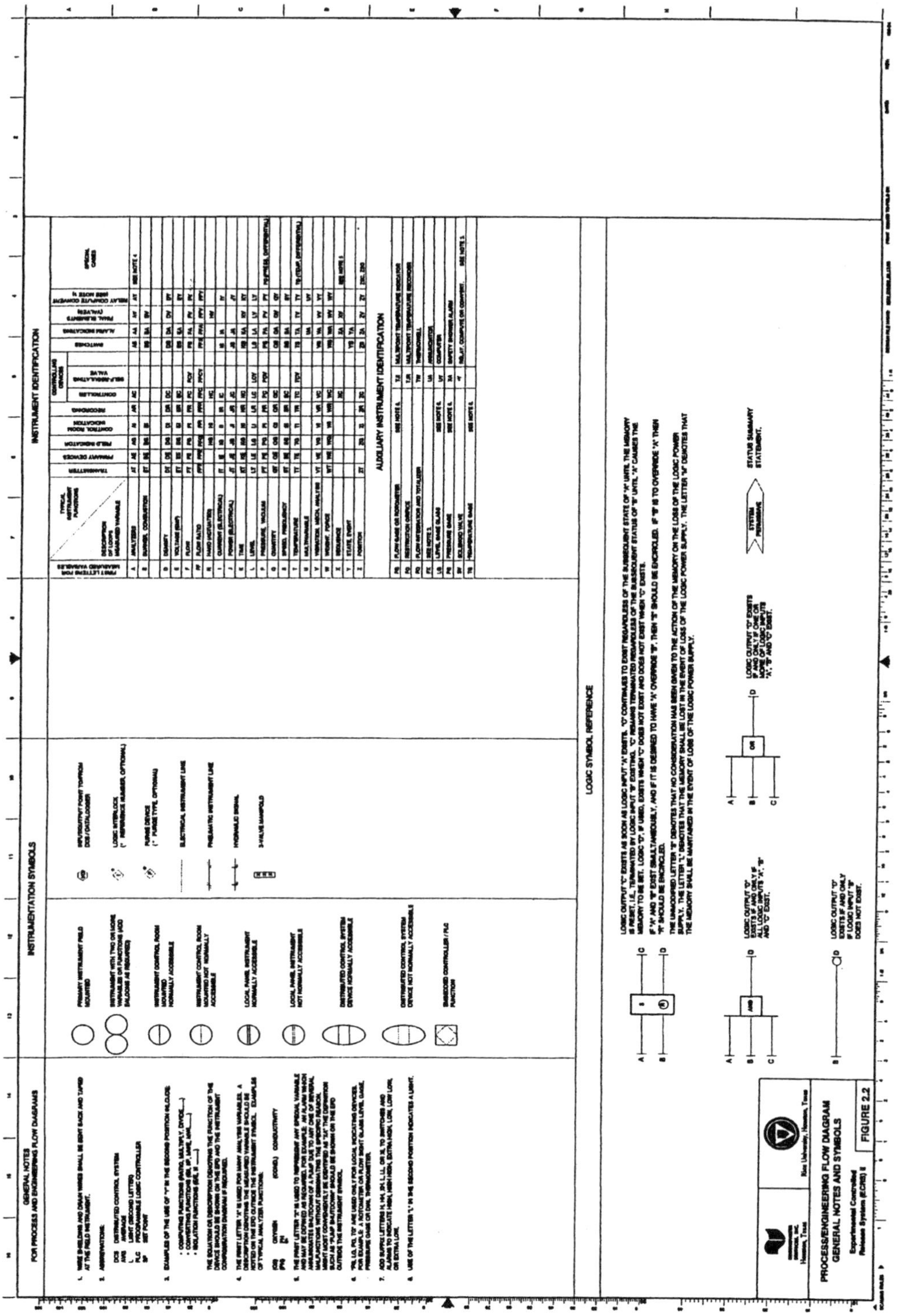
GENERAL NOTES FOR PROCESS AND ENGINEERING FLOW DIAGRAMS
INSTRUMENTATION SYMBOLS
INSTRUMENT IDENTIFICATION
AUXILIARY INSTRUMENT IDENTIFICATION
LOGIC SYMBOL REFERENCE
PROCESS/ENGINEERING FLOW DIAGRAM
GENERAL NOTES AND SYMBOLS

FIGURE 2.2

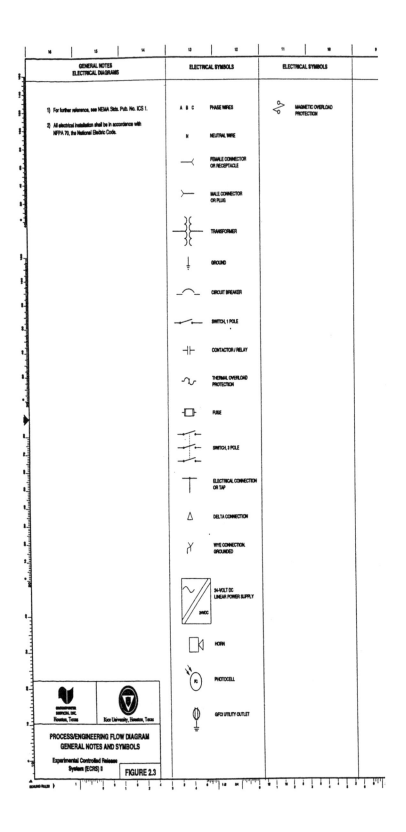

FIGURE 2.3

PROCESS/ENGINEERING FLOW DIAGRAM
GENERAL NOTES AND SYMBOLS

Experimental Controlled Release System (ECRS) II

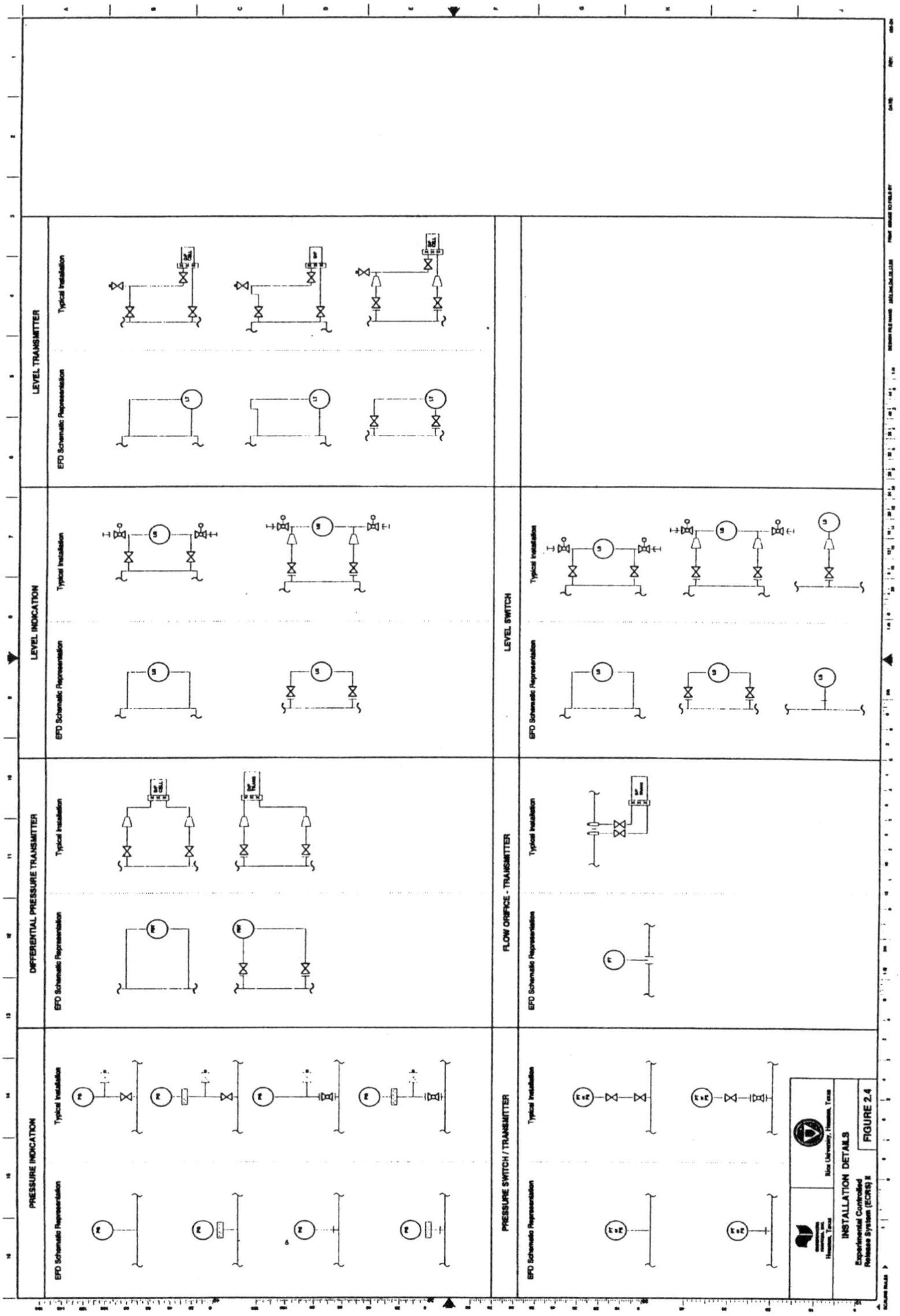
PRESSURE INDICATION
DIFFERENTIAL PRESSURE TRANSMITTER
LEVEL INDICATION
LEVEL TRANSMITTER
PRESSURE SWITCH / TRANSMITTER
FLOW ORIFICE - TRANSMITTER
LEVEL SWITCH
EFD Schematic Representation
Typical Installation
INSTALLATION DETAILS
Experimental Controlled Release System (ECRS) II
Rice University, Houston, Texas
FIGURE 2.4

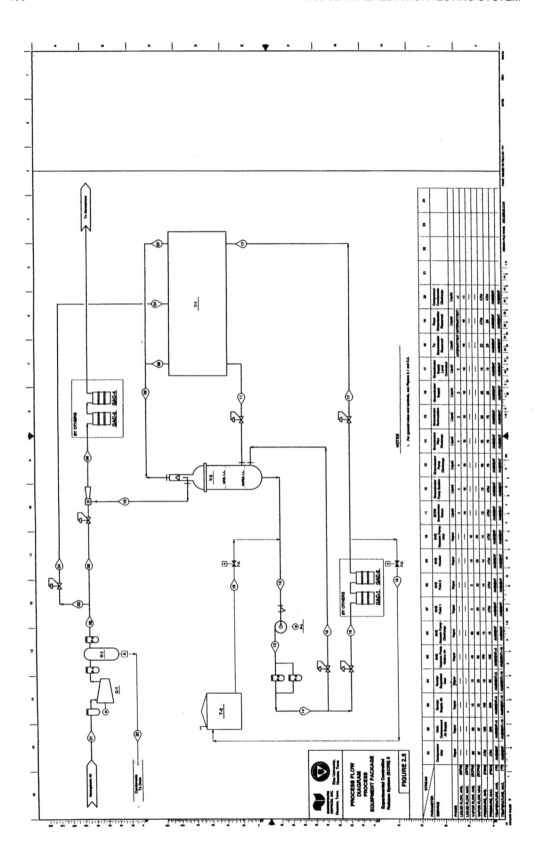

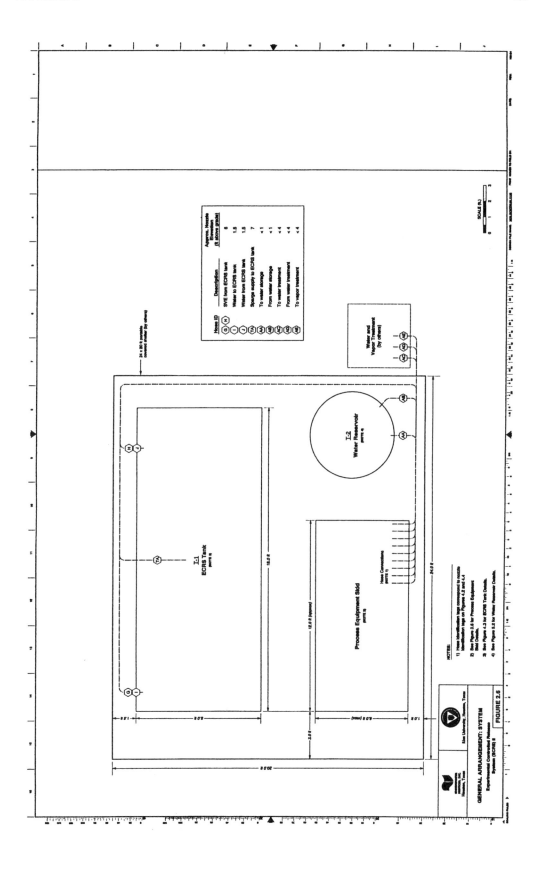

GSI Job No. G-1874
September 13, 1996

ENGINEERING SPECIFICATIONS

Experimental Controlled Release System (ECRS) II
DOD-AATDF/Rice University, Houston, Texas

SECTION 3.0 SPECIFICATIONS

SECTION 3.0 FIGURES

GSI Job No. G-1874
September 13, 1996

(see Figure 3.4). Additional system controls and electrical requirements are provided on Figures 3.3 - 3.5.

Page 1 of 1

SPECIFICATION 3.1
EQUIPMENT SPECIFICATION: AIR COMPRESSOR (C-1)

An Ingersoll-Rand electric screw air compressor Model SSR 15-50 HP SE shall be provided. The compressor model shall meet the Ingersoll-Rand specifications Reference Nos. 7250.02, 7252.01, 7252.02, 7252.03 dated June 1, 1994, and 7254.09A, 7254.09 dated June 15, 1995.

GSI Job No. G-1874
Issued: 9/13/96
Page 1 of 1

SPECIFICATION 3.2
EQUIPMENT SPECIFICATION:
AIR FILTERS

Experimental Controlled Release System (ECRS) II
Rice University, Houston, Texas

Tag No.			F-1		F-2	
Service:			Compressed Air		Compressed Air	
EFD No.			Figure 3.1		Figure 3.1	
Line No.			Air Compressor Discharge		Discharge from Air Accumulator	
Process Conditions						
Flowrate (scfm)						
Average		Maximum	80	87	80	87
Pressure (psig)						
Average		Maximum	150	185	150	185
Temperature (°F)						
Average		Maximum	70	100	70	100
Filter Specifications						
Type			Coalescing		Cartridge	
Size					9.75 in long	
Materials of Construction						
Housing					Nickel Plated Brass & Stainless Steel	
Gasket					Nitrile	
Cartridge					Extruded carbon media	
Connection					1 in FNPT	
Drain						
Size (in)					1/4	
Type					Off-set	
Purchasing						
Manufacturer			See attached specification from Ingersoll-Rand Ref: 11502.01 3/15/93		Cuno Inc. Process Filtration Products 400 Research Parkway Meridian, CT 06450 (203) 237-541	
Model			IR125C		CT101A-47310-01 (housing) CT0-03-250-125-975 (cartridge) 3551-01 (mounting bracket)	
Notes:						

Rev	Date	By	Ch'd	App'd	Description
0	9/13/96	EAH	TLR		Release for Bid

GSI Job No. G-1874
Issued: 9/13/96
Page 1 of 1

SPECIFICATION 3.3
EQUIPMENT SPECIFICATION: AIR RECEIVER (V-3)

An 80 gallon air receiver vessel with a working pressure of 200 psi shall be provided. The vessel shall be mounted in a vertical position with suitable mounting arrangement (i.e. mounting feet or a suitable vessel skirt.) The maximum allowable outside diameter of the vessel shall be 24 inches, nominal. Appropriate vent, drain, inspection, and process connections shall be provided. An automatic drain with trap shall be installed at low point of vessel.

The air receiver vessel shall be designed and constructed in accordance with the latest American Society of Mechanical Engineers boiler and pressure vessel code and shall bear an identification plate indicating conformance with this code. Materials of construction shall be carbon steel with exterior primer coating and suitable interior corrosion resistant coating.

GSI Job No. G-1874
Issued: 9/13/96
Page 1 of 1

SPECIFICATION 3.4
EQUIPMENT SPECIFICATION: VENTURI VACUUM PUMPS

Experimental Controlled Release System (ECRS) II
Rice University, Houston, Texas

Tag No.	SP-1	SP-2
Service:	Air	Air
EFD No.	Figure 3.1	Figure 3.1
Line No.	Applied SVE Vacuum	Reference SVE Vacuum
Unit Specifications		
Type	Venturi	Venturi
Dimensions (L x W x H)	See attached drawing 613667	1" high x 1/2" wide x 1-5/8" long
Materials of Construction	304 stainless steel	Manufacturer Standard
Connections		
Motive (in)	3/4 in FNPT	5/32 in
Suction (in)	1-1/2 in MNPT	5/32 in
Discharge (in)	1-1/2 in MNPT	5/32 in
Motive	45 scfm @ 125 psi	0.77 cfm @ 120 psi
Suction	28 in Hg abs., 30 scfm air @ 70°F	10 in Hg abs., 0.16 cfm @ 70°F
Discharge	18 psia	15.2 psia
Purchasing		
Manufacturer/Supplier	Fox Valve Development Corp. Hamilton Business Park, Unit 6A Dover, NJ 07801 (201) 328-1011	Crouzet Corp. 2445 Midway Road Carrollton, TX 75006-2503
Model	Series 250-AJV	81 45 001

Notes:
Provide silencer for SP-2 discharge, min CV = 4.0.

Rev	Date	By	Ck'd	App'd	Description
0	9/13/96	EAH	TLR		Issue for bid

GSI Job No. G-1874
Issued: 9/13/96
Page 1 of 1

SPECIFICATION 3.5
EQUIPMENT SPECIFICATION: PUMPS

Experimental Controlled Release System (ECRS) II
Rice University, Houston, Texas

Tag No.			P-1			P-2		
Service:			Water			Water		
EFD No.			Figure 3.1			Figure 3.1		
Line No.			Water circulation			Utility water		
Process Conditions								
Flowrate (gpm)								
Max	Norm	Min	10	6	2	20	5	2
Temperature (°F)								
Max	Norm	Min	100	70	40	100	70	40
Fluid								
Sp. Gr.			1			1		
Viscosity			1 cp			1 cp		
Duty Cycle			Continuous			Intermittent		
Service (%)			100			10		
Moisture (%)			100			100		
Abrasives			Light			Light		
Solids								
Fraction (%)			<1			<1		
Size (μm)			<200 (fine silt)			<200 (fine silt)		
Unit Specifications								
Type			Centrifugal-multistage			Centrifugal		
Dimensions (L x W x H)			9-7/8 in x 8-1/4 in x20-3/4			13.5 in x 7 in x 8.5 in		
Materials of Construction			SS			316 SS		
Gaskets/Seals			Viton			Teflon		
Connections								
Inlet (in)			1			1		
Outlet (in)			1			1		
Discharge Pressure (ft H$_2$O)								
Max	Norm	Min	150	135	120	24	22	20
Power Requirements			480 V ac, 3 phase			115/230 V ac, single phase		
Purchasing								
Manufacturer			ITT Bell & Gossett Morton Grove, IL 60053 (708) 966-3700			Price Pump Co. Sonoma, CA95476 (707) 938-8441		
Model			3540-20-3			CD100SS-494-21211-PEQ		

Notes:
Utility water pump to be provided to contractor by Rice University.

Rev	Date	By	Chk'd	App'd	Description
A	8/23/96	RLB	EAH		Preliminary for review
0	9/13/96	RLB	EAH		Issue for bid

GSI Job No. G-1874
Issued: 9/13/96
Page 1 of 1

SPECIFICATION 3.6
EQUIPMENT SPECIFICATION:
WATER FILTERS

Experimental Controlled Release System (ECRS) II
Rice University, Houston, Texas

Tag No.		F-3		F-4	
Service:		Water		Water	
EFD No.		Figure 3.1		Figure 3.1	
Line No.		Groundwater Pump Discharge		Groundwater Pump Discharge	
Process Conditions					
Flowrate (gpm)					
Average	Maximum	6	10	6	10
Pressure (psig)					
Average	Maximum	55	75	55	75
Temperature (°F)					
Average	Maximum	70	100	70	100
Fluid					
Sp. Gr.	Density (lb/ft³)	1		1	
Vacuum Possibility		No		No	
Filter Specifications					
Type		Spiral Wound		Spiral Wound	
Line Size (in)		1		1	
Materials of Construction					
Housing		Brass and Stainless Steel		Brass and Stainless Steel	
Cartridge					
Connections					
Connection Type		NPT		NPT	
Purchasing					
Manufacturer		Cuno Inc.		Cuno Inc.	
		Process Filtration Products		Process Filtration Products	
		400 Research Parkway		400 Research Parkway	
		Meridian, CT 06450		Meridian, CT 06450	
		(203) 237-5541		(203) 237-5541	
Model		CT101		CT101	
Notes:					

Rev	Date	By	Ch'd	App'd	Description
A	8/23/96	EAH			Release for Review
0	9/13/96	EAH			Release for Bid

GSI Job No. G-1874
Issued: 9/13/96
Page 1 of 2

SPECIFICATION 3.7
INSTRUMENTATION SPECIFICATION:
SOLENOID VALVE

Experimental Controlled Release System (ECRS) II
Rice University, Houston, Texas

GENERAL			XV-1	XV-2	XV-3
	1	Tag No.	XV-1	XV-2	XV-3
	2	Service	Air	Air	Water
	3	Line No./Vessel No.	Compressed Air Supply	Sparge Supply	GW Reservoir Return
	4	Quantity	1	1	1
VALVE BODY	5	Type	2-way	2-way	2-way
	6	Size—Body/Port	1 / 1-5/8	1 / 1-5/8	1 / 1-5/8
	7	Rating & Type Conn	150# NPT	150# NPT	150# NPT
	8	Material—Body	Aluminum	Aluminum	Aluminum
	9	Material—Seat	Buna-n	Buna-n	Buna-n
	10	Material—Diaphragm	Buna-n	Buna-n	Buna-n
	11	Operation Direct/Pilot	Direct	Direct	Direct
	12	Packless or Type Packed			
	13	Manual Re-Set			
	14	Manual Operator			
WHEN DE-ENERGIZED	15	2-Way Valve Opens/Closes	NC	NC	NC
	16	3-Way			
	17	Vent Port Opens/Closes			
	18	Press Port Opens/Closes			
	19	4-Way			
	20	Press to Cyl. 1/Cyl. 2			
	21	Exh. from Cyl. 1/Cyl. 2			
SOLENOID	22	Enclosure	NEMA 4, 7	NEMA 4, 7	NEMA 4, 7
	23	Voltage/Hz	120 V, 60 Hz	120 V, 60 Hz	120 V, 60 Hz
	24	Style of Coil	F	F	F
	25	Single or Double Coil	Single	Single	Single
	26	VA Inrush/Hold	160/27	160/27	160/27
	27	Area Classification	Class 1, Division 2	Class 1, Division 2	Class 1, Division 2
SERVICE CONDITIONS	28	Fluid	Air	Air	Water
	29	Qty. Maximum	90 SCFM	30 SCFM	10 gpm
	30	Oper. Diff. Min/Max PSI	0 / 1.0	0 / 0.1	0 / 0.1
	31	Allow. Diff. Min/Max			
	32	Temp. Norm/Max °F	75 / 110	75 / 110	75 / 90
	33	Oper. sp. gr.	1	1	1
	34	Oper Viscosity	0.018 cp	0.018 cp	1 cp
	35	Required Cv	11	4	10
	36	Valve Cv	21	21	21
	37	Manufacturer	ASCO	ASCO	ASCO
	38	Model No.	EF 8215B50	EF 8215B50	EF 8215B50

NOTES
1. NA = Not applicable.

GSI Job No. G-1874
Issued: 9/13/96
Page 2 of 2

SPECIFICATION 3.7
INSTRUMENTATION SPECIFICATION:
SOLENOID VALVE

Experimental Controlled Release System (ECRS) II
Rice University, Houston, Texas

GENERAL	1	Tag No.	XV-4	XV-5	XV-6
	2	Service	Water	Water	Water
	3	Line No./Vessel No.	GW Return to ECRS Tank	GW Reservoir Supply	Groundwater Supply
	4	Quantity	1	1	1
VALVE BODY	5	Type	2-way	2-way	2-way
	6	Size—Body/Port	2 / 2-3/32	1 / 1-5/8	1 / 1-5/8
	7	Rating & Type Conn	150# NPT	150# NPT	150# NPT
	8	Material—Body	Aluminum	Aluminum	Aluminum
	9	Material—Seat	Buna-n	Buna-n	Buna-n
	10	Material—Diaphragm	Buna-n	Buna-n	Buna-n
	11	Operation Direct/Pilot	Direct	Direct	Direct
	12	Packless or Type Packed			
	13	Manual Re-Set			
	14	Manual Operator			
WHEN	15	2-Way Valve Opens/Closes	NC	NC	NC
DE-ENERGIZED	16	3-Way			
	17	Vent Port Opens/Closes			
	18	Press Port Opens/Closes			
	19	4-Way			
	20	Press to Cyl. 1/Cyl. 2			
	21	Exh. from Cyl. 1/Cyl. 2			
SOLENOID	22	Enclosure	NEMA 4, 7	NEMA 4, 7	NEMA 4, 7
	23	Voltage/Hz	120 V, 60 Hz	120 V, 60 Hz	120 V, 60 Hz
	24	Style of Coil	F	F	F
	25	Single or Double Coil	Single	Single	Single
	26	VA Inrush/Hold	160/27	160/27	160/27
	27	Area Classification	Class 1, Division 2	Class 1, Division 2	Class 1, Division 2
SERVICE	28	Fluid	Water	Water	Water
CONDITIONS	29	Qty. Maximum	10 gpm	10 gpm	10 gpm
	30	Oper. Diff. Min/Max PSI	0 / 0.1	0 / 0.1	0 / 0.1
	31	Allow. Diff. Min/Max			
	32	Temp. Norm/Max °F	75 / 90	75 / 90	75 / 90
	33	Oper. sp. gr.	1	1	1
	34	Oper Viscosity	1 cp	1 cp	1 cp
	35	Required Cv	52.6	10	52.6
	36	Valve Cv	60	21	60
	37	Manufacturer	ASCO	ASCO	ASCO
	38	Model No.	EF 8215B80	EF 8215B50	EF 8215B50

NOTES
1. NA = Not applicable.

GSI Job No. 1874-1
Issued: 9/13/96
Page 1 of 6

SPECIFICATION 3.8
PRESSURE REGULATORS

Experimental Controlled Release System (ECRS) II
Rice University, Houston, Texas

GENERAL			BPV-1		BPV-2	
1	Tag No.		BPV-1		BPV-2	
2	Service		GW RETURN HEAD		GW FLOW BYPASS	
3	Line No./Vessel No.					
4	Line Size/Schedule		1 1/2 SCH 40		1 SCH 40	
5	Function		BACKPRESSURE CONTROL		BACKPRESSURE CONTROL	
6	Actuation		DOME LOAD		SPRING	
BODY 7	Type of Body		MFR. STD.		MFR. STD.	
8	Body Size	Port Size				
9	Guiding	No. of Ports	MFR. STD.		MFR. STD.	
10	End Conn. & Rating		1 1/2 NPT		1 NPT	
11	Body Material		DUCTILE IRON		DUCTILE IRON	
12	Packing Material					
13	Lubricator	Iso. Valve				
14	Seal Type					
15	Trim Form					
16	Trim Material					
17	Seat Material		VITON		VITON	
18	Required Seat Tightness					
19	Max Allow Sound Level dBA		85		85	
ACTUATOR/ 20	Type of Actuator		DIAPHRAGM		DIAPHRAGM	
PILOT 21	Pilot					
22	Supply to Pilot					
23	Self Contained	Ext. Conn.		EXTERNAL		EXTERNAL
24	Diaphragm Material		VITON		VITON	
25	Diaphragm Rating					
26	Spring Range		N/A		0 - 75 PSI	
27	Set Point				60 PSI	
28	Regulation Accuracy		± 1/2 INCH WATER		± 2 PSI	
ACCESSORIES 29	Filt. Reg.	Supply Gage				
30	Line Strainer					
31	Housing Vent					
32	Internal Relief					
33	Mounting		PIPE		PIPE	
SERVICE 34	FLOW UNITS		LIQUID, GPM		LIQUID, GPM	
35	Fluid		WATER		WATER	
36	Quant. Max	Cv	10		10	
37	Quant. Oper	Cv	4		4	
38	Valve Cv	Valve FL				
39	Norm. Inlet Press.	dP	39 IN WATER	± 27 IN WATER	60 PSI	± 5 PSI
40	Max. Inlet Pressure		66 IN WATER		65 PSI	
41	Max Shut Off	dP	72 IN WATER		65 PSI	
42	Temp Max	Operating	90 °F	80 °F	90 °F	80 °F
43	Oper. Sp. Gr.	Mol. Wt.	1		1	
44	Oper. Visc.	% Flash				
45	% Superheat	% Solids		~ 1 %	~ 1 %	
46	Vapor Press.	Crit. Press.				
47	Predicted Sound Level dBA					
48	Manufacturer		KAYE & MACDONALD		KAYE & MACDONALD	
49	Model No.					

NOTES

GSI Job No. 1874-1
Issued: 9/13/96
Page 2 of 6

SPECIFICATION 3.8
PRESSURE REGULATORS

Experimental Controlled Release System (ECRS) II
Rice University, Houston, Texas

				PRV-1		PRV-2	
GENERAL	1	Tag No.		PRV-1		PRV-2	
	2	Service		REF GW SUPPLY HEAD		REF GW RETURN HEAD	
	3	Line No./Vessel No.					
	4	Line Size/Schedule		3/8 SCH 40		3/8 SCH 40	
	5	Function		PRESSURE REDUCING		PRESSURE REDUCING	
	6	Actuation		SPRING		SPRING	
BODY	7	Type of Body		MFR. STD.		MFR. STD.	
	8	Body Size	Port Size				
	9	Guiding	No. of Ports	MFR. STD.		MFR. STD.	
	10	End Conn. & Rating		3/8 NPT		3/8 NPT	
	11	Body Material		DIECAST ALUMINUM		DIECAST ALUMINUM	
	12	Packing Material					
	13	Lubricator	Iso. Valve				
	14	Seal Type					
	15	Trim Form					
	16	Trim Material					
	17	Seat Material					
	18	Required Seat Tightness					
	19	Max Allow Sound Level dBA		85		85	
ACTUATOR/ PILOT	20	Type of Actuator		ROLLING DIAPHRAGM		ROLLING DIAPHRAGM	
	21	Pilot					
	22	Supply to Pilot					
	23	Self Contained	Ext. Conn.	SELF		SELF	
	24	Diaphragm Material		NITRILE & POLYESTER		NITRILE & POLYESTER	
	25	Diaphragm Rating					
	26	Spring Range		0 - 15 PSI		0 - 15 PSI	
	27	Set Point					
	28	Regulation Accuracy		± 1/2 INCH WATER		± 1/2 INCH WATER	
ACCESSORIES	29	Filt. Reg.	Supply Gage				
	30	Line Strainer					
	31	Housing Vent					
	32	Internal Relief					
	33	Mounting		PANEL		PANEL	
SERVICE	34	FLOW UNITS		GAS, SCFM		GAS, SCFM	
	35	Fluid		AIR		AIR	
	36	Quant. Max	Cv	5		5	
	37	Quant. Oper	Cv	<1		<1	
	38	Valve Cv	Valve FL				
	39	Norm. Inlet Press.	dP	150 PSI	± 2 PSI	150 PSI	± 2 PSI
	40	Max. Inlet Pressure		165 PSI		165 PSI	
	41	Max Shut Off	dP	165 PSI		165 PSI	
	42	Temp Max	Operating				
	43	Oper. Sp. Gr.	Mol. Wt.	1		1	
	44	Oper. Visc.	% Flash		0		0
	45	% Superheat	% Solids	0	0	0	0
	46	Vapor Press.	Crit. Press.				
	47	Predicted Sound Level dBA					
	48	Manufacturer					
	49	Model No.					

NOTES

GSI Job No. 1874-1
Issued: 9/13/96
Page 3 of 6

SPECIFICATION 3.8
PRESSURE REGULATORS

Experimental Controlled Release System (ECRS) II
Rice University, Houston, Texas

GENERAL	1	Tag No.		PRV-3		PRV-4	
	2	Service		AIR SPARGE		REF SVE VACUUM	
	3	Line No./Vessel No.					
	4	Line Size/Schedule		1 SCH 40		3/8 SCH 40	
	5	Function		PRESSURE REDUCING		PRESSURE REDUCING	
	6	Actuation		DOME LOAD		SPRING	
BODY	7	Type of Body		MFR. STD.		MFR. STD.	
	8	Body Size	Port Size				
	9	Guiding	No. of Ports	MFR. STD.		MFR. STD.	
	10	End Conn. & Rating		1 NPT		3/8 NPT	
	11	Body Material		DUCTILE IRON		DIECAST ALUMINUM	
	12	Packing Material					
	13	Lubricator	Iso. Valve				
	14	Seal Type					
	15	Trim Form					
	16	Trim Material					
	17	Seat Material					
	18	Required Seat Tightness					
	19	Max Allow Sound Level dBA		85		85	
ACTUATOR/ PILOT	20	Type of Actuator		ROLLING DIAPHRAGM		ROLLING DIAPHRAGM	
	21	Pilot					
	22	Supply to Pilot					
	23	Self Contained	Ext. Conn.		EXTERNAL	SELF	
	24	Diaphragm Material		NITRILE & POLYESTER		NITRILE & POLYESTER	
	25	Diaphragm Rating					
	26	Spring Range		0 - 15 PSI		2 - 150 PSI	
	27	Set Point					
	28	Regulation Accuracy		± 2 INCH WATER		± 1 PSI	
ACCESSORIES	29	Filt. Reg.	Supply Gage				
	30	Line Strainer					
	31	Housing Vent					
	32	Internal Relief					
	33	Mounting		PIPE		PANEL	
SERVICE	34	FLOW UNITS		GAS, SCFM		GAS, SCFM	
	35	Fluid		AIR		AIR	
	36	Quant. Max	Cv	30		10	
	37	Quant. Oper	Cv	20		5	
	38	Valve Cv	Valve FL				
	39	Norm. Inlet Press.	dP	150 PSI	± 2 PSI	150 PSI	± 2 PSI
	40	Max. Inlet Pressure		165 PSI		165 PSI	
	41	Max Shut Off	dP	165 PSI		165 PSI	
	42	Temp Max	Operating				
	43	Oper. Sp. Gr.	Mol. Wt.	1		1	
	44	Oper. Visc.	% Flash		0		0
	45	% Superheat	% Solids	0	0	0	0
	46	Vapor Press.	Crit. Press.				
	47	Predicted Sound Level dBA					
	48	Manufacturer					
	49	Model No.					

NOTES

GSI Job No. 1874-1
Issued: 9/13/96
Page 4 of 6

SPECIFICATION 3.8
PRESSURE REGULATORS

Experimental Controlled Release System (ECRS) II
Rice University, Houston, Texas

GENERAL	1	Tag No.		PRV-5		PRV-6	
	2	Service		UTILITY AIR		APPLIED VACUUM	
	3	Line No./Vessel No.					
	4	Line Size/Schedule		3/8 SCH 40		3/4 SCH 40	
	5	Function		PRESSURE REDUCING		PRESSURE REDUCING	
	6	Actuation		SPRING		DOME LOAD	
BODY	7	Type of Body		MFR. STD.		MFR. STD.	
	8	Body Size	Port Size				
	9	Guiding	No. of Ports	MFR. STD.		MFR. STD.	
	10	End Conn. & Rating		3/8 NPT		3/4 NPT	
	11	Body Material		DIECAST ALUMINUM		DUCTILE IRON	
	12	Packing Material					
	13	Lubricator	Iso. Valve				
	14	Seal Type					
	15	Trim Form					
	16	Trim Material					
	17	Seat Material				VITON	
	18	Required Seat Tightness					
	19	Max Allow Sound Level dBA		85		85	
ACTUATOR/ PILOT	20	Type of Actuator		ROLLING DIAPHRAGM		DIAPHRAGM	
	21	Pilot					
	22	Supply to Pilot					
	23	Self Contained	Ext. Conn.	SELF			EXTERNAL
	24	Diaphragm Material		NITRILE & POLYESTER		VITON	
	25	Diaphragm Rating					
	26	Spring Range		2 - 150 PSI		N/A	
	27	Set Point					
	28	Regulation Accuracy		± 2 PSI		± 1 PSI	
ACCESSORIES	29	Filt. Reg.	Supply Gage				
	30	Line Strainer					
	31	Housing Vent					
	32	Internal Relief					
	33	Mounting		PANEL		PIPE	
SERVICE	34	FLOW UNITS		GAS, SCFM		GAS, SCFM	
	35	Fluid		AIR		AIR	
	36	Quant. Max	Cv	30		60	
	37	Quant. Oper	Cv	15		45	
	38	Valve Cv	Valve FL				
	39	Norm. Inlet Press.	dP	150 PPSI	± 2 PSI	150 PSI	± 2 PSI
	40	Max. Inlet Pressure		165 PSI		165 PSI	
	41	Max Shut Off	dP	165 PSI		165 PSI	
	42	Temp Max	Operating				
	43	Oper. Sp. Gr.	Mol. Wt.	1		1	
	44	Oper. Visc.	% Flash		0		
	45	% Superheat	% Solids	0	0		
	46	Vapor Press.	Crit. Press.				
	47	Predicted Sound Level dBA					
	48	Manufacturer				KAYE & MACDONALD	
	49	Model No.					

NOTES

GSI Job No. 1874-1
Issued: 9/13/96
Page 5 of 6

SPECIFICATION 3.8
PRESSURE REGULATORS

Experimental Controlled Release System (ECRS) II
Rice University, Houston, Texas

GENERAL	1	Tag No.		PRV-7		PRV-8
	2	Service		REF AIR SPARGE		GW SUPPLY HEAD
	3	Line No./Vessel No.				
	4	Line Size/Schedule		3/8 SCH 40		1 SCH 40
	5	Function		DIFFERENTIAL PRESSURE		PRESSURE REDUCING
	6	Actuation		SPRING		DOME LOAD
BODY	7	Type of Body		MFR. STD.		MFR. STD.
	8	Body Size	Port Size			
	9	Guiding	No. of Ports	MFR. STD.		MFR. STD.
	10	End Conn. & Rating		3/8 NPT		1 NPT
	11	Body Material				DUCTILE IRON
	12	Packing Material				
	13	Lubricator	Iso. Valve			
	14	Seal Type				
	15	Trim Form				
	16	Trim Material				
	17	Seat Material				VITON
	18	Required Seat Tightness				
	19	Max Allow Sound Level dBA		85		85
ACTUATOR/	20	Type of Actuator		ROLLING DIAPHRAGM		DIAPHRAGM
PILOT	21	Pilot				
	22	Supply to Pilot				
	23	Self Contained	Ext. Conn.		EXTERNAL		EXTERNAL
	24	Diaphragm Material		NITRILE & POLYESTER		VITON
	25	Diaphragm Rating				
	26	Spring Range		0 - 15 PSI		N/A
	27	Set Point				
	28	Regulation Accuracy		± 1/2 INCH WATER		± 1/2 INCH WATER
ACCESSORIES	29	Filt. Reg.	Supply Gage			
	30	Line Strainer				
	31	Housing Vent				
	32	Internal Relief				
	33	Mounting		PANEL		PIPE
SERVICE	34	FLOW UNITS		GAS, SCFM		LIQUID, GPM
	35	Fluid		AIR		WATER
	36	Quant. Max	Cv	5		10
	37	Quant. Oper	Cv	<1		4
	38	Valve Cv	Valve FL			
	39	Norm. Inlet Press.	dP	150 PSI	± 2 PSI	9 PSI	± 1/2 PSI
	40	Max. Inlet Pressure		165 PSI		10 PSI
	41	Max Shut Off	dP	165 PSI		65 PSI
	42	Temp Max	Operating			90 °F	80 °F
	43	Oper. Sp. Gr.	Mol. Wt.	1		1
	44	Oper. Visc.	% Flash		0	
	45	% Superheat	% Solids	0	0	
	46	Vapor Press.	Crit. Press.			
	47	Predicted Sound Level dBA				
	48	Manufacturer				KAYE & MACDONALD
	49	Model No.				

NOTES

GSI Job No. 1874-1
Issued: 9/13/96
Page 6 of 6

SPECIFICATION 3.8
PRESSURE REGULATORS

Experimental Controlled Release System (ECRS) II
Rice University, Houston, Texas

GENERAL	1	Tag No.		PRV-9			
	2	Service		GW TREATMENT SUPPLY			
	3	Line No./Vessel No.					
	4	Line Size/Schedule		1 SCH 40			
	5	Function		PRESSURE REDUCING			
	6	Actuation		SPRING			
BODY	7	Type of Body		MFR. STD.			
	8	Body Size	Port Size				
	9	Guiding	No. of Ports	MFR. STD.			
	10	End Conn. & Rating		1 NPT			
	11	Body Material		DUCTILE IRON			
	12	Packing Material					
	13	Lubricator	Iso. Valve				
	14	Seal Type					
	15	Trim Form					
	16	Trim Material					
	17	Seat Material		VITON			
	18	Required Seat Tightness					
	19	Max Allow Sound Level dBA		85			
ACTUATOR/	20	Type of Actuator		DIAPHRAGM			
PILOT	21	Pilot					
	22	Supply to Pilot					
	23	Self Contained	Ext. Conn.	SELF			
	24	Diaphragm Material		VITON			
	25	Diaphragm Rating					
	26	Spring Range		N/A			
	27	Set Point		9 PSI			
	28	Regulation Accuracy		± 1/2 PSI			
ACCESSORIES	29	Filt. Reg.	Supply Gage	•			
	30	Line Strainer					
	31	Housing Vent					
	32	Internal Relief					
	33	Mounting		PIPE			
SERVICE	34	FLOW UNITS		LIQUID, GPM			
	35	Fluid		WATER			
	36	Quant. Max	Cv	10			
	37	Quant. Oper	Cv	4			
	38	Valve Cv	Valve FL				
	39	Norm. Inlet Press.	dP	60	± 2 PSI		
	40	Max. Inlet Pressure		65 PSI			
	41	Max Shut Off	dP	65 PSI			
	42	Temp Max	Operating	90 °F	80 °F		
	43	Oper. Sp. Gr.	Mol. Wt.	1			
	44	Oper. Visc.	% Flash				
	45	% Superheat	% Solids				
	46	Vapor Press.	Crit. Press.				
	47	Predicted Sound Level dBA					
	48	Manufacturer		KAYE & MACDONALD			
	49	Model No.					

NOTES

GSI Job No. G-1874
Issued: 9/13/96
Page 1 of 1

SPECIFICATION 3.9
INSTRUMENTATION SPECIFICATION
DATA RECORDER (UIR-1)

Data recorder shall be a 250-mm strip chart recorder with memory card archiving of all data in ASCII format. The recorder shall have the capacity to trace 45 variables in 6 colors. Local display of up to 4 variables shall be 80 character, 3-color vacuum fluorescent display. The recorder shall be capable of calculating derived variables (i.e. mass flow from volume flow, pressure, and temperature.) Data recorder shall have up to 4 alarms available per channel. These alarms may be: absolute hi/lo, deviation, rate-of-change increasing/decreasing, or digital status. Data recorder shall be Eurotherm Chessel Model 4250M configured with 4 8-channel universal input cards, 2 6-channel relay output cards and the following options.

Power:	90-132 VAC
Printing:	14 pin dot matrix with 6 color ribbon
Chart:	250-mm, Z-fold, 22-m
Display:	80 char., 4 variable, vacuum fluorescent
Inputs:	4 8-channel universal input cards
Outputs:	2 6-channel relay output cards
Options:	CEM Package: Math functions; 12 each counters, timers and totalizers; 96 derived variable calculations; 2048k memory card for data archiving

GSI Job No. G-1874
Issued: 9/13/96
Page 1 of 3

SPECIFICATION 3.10
INSTRUMENTATION SPECIFICATION:
PRESSURE TRANSMITTER

Experimental Controlled Release System (ECRS) II
Rice University, Houston, Texas

Tag No.		PIT-1		PIT-2	
Service:		Atmospheric air		Extracted soil vapor	
EFD No.		Figure 3.1		Figure 3.1	
Line No.		Sparge Mass Flow		SVE Mass Flow	
Process Conditions					
Flowrate (scfm)					
Average	Max	15	30	80	87
Pressure (psig)					
Average	Max	4	7	1 (vacuum)	Atm
Temperature (°F)					
Average	Max	70	90	70	90
Vacuum Possibility		No		Yes	
Meter Specifications					
Function		Indicate & transmit pressure		Indicate & transmit pressure	
Display		Integral		Integral	
Materials of Construction					
Enclosure					
Process Seal Gasket		Viton		Viton	
Process Connection		Stainless steel		Stainless steel	
Sensor					
Diaphragm seal required?		No		No	
Conduit Entry					
Mounting		Surface, 1/2 in MNPT		Surface, 1/2 in MNPT	
Electrical					
Power Supply					
Grounding					
Electrical Classification		Not classified		Not classified	
Enclosure Classification					
Range		0 - 5 psig		0 - 10 psig (vacuum)	
Accuracy		Manufacturer standard		Manufacturer standard	
Output		4 - 20 mA		4 - 20 mA	
Display		LCD, units of in H_2O		LCD, units of in H_2O	
Purchasing					
Manufacturer		Endress + Hauser 600 Kenrick, Suite B-3 Houston, Texas 77060 (713) 999-1991		Endress + Hauser 600 Kenrick, Suite B-3 Houston, Texas 77060 (713) 999-1991	
Model		Cerabar S 430		Cerabar S 430	
Notes:					

Rev	Date	By	Ck'd	App'd	Description
A	8/23/96	EAH			Issued for Review
0	9/13/96	EAH			Issued for Bid

GSI Job No. G-1874
Issued: 9/13/96
Page 2 of 3

SPECIFICATION 3.10
INSTRUMENTATION SPECIFICATION:
PRESSURE TRANSMITTER

Experimental Controlled Release System (ECRS) II
Rice University, Houston, Texas

Tag No.			PIT-3		PIT-4	
Service:			Extracted soil vapor		Atmospheric Air	
EFD No.			Figure 3.1		Figure 3.1	
Line No.			SVE Pressure Sense		Sparge Supply	
Process Conditions						
Flowrate (scfm)						
Average	Max		80	87	15	30
Pressure (psig)						
Average	Max		1 (vacuum)	Atm	4	7
Temperature (°F)						
Average	Max		70	90	70	90
Vacuum Possibility			Yes		No	
Meter Specifications						
Function			Indicate & transmit pressure		Indicate & transmit pressure	
Display			Integral		Integral	
Materials of Construction						
Enclosure						
Process Seal Gasket			Viton		Viton	
Process Connection			Stainless steel		Stainless steel	
Sensor						
Diaphragm seal required?			No		No	
Conduit Entry						
Mounting			Surface, 1/2 in MNPT		Surface, 1/2 in MNPT	
Electrical						
Power Supply						
Grounding						
Electrical Classification			Not classified		Not classified	
Enclosure Classification						
Range			0 - 10 psig (vacuum)		0 - 5 psig	
Accuracy			Manufacturer standard		Manufacturer standard	
Output			4 - 20 mA		4 - 20 mA	
Display			LCD, units of in H_2O		LCD, units of in H_2O	
Purchasing						
Manufacturer			Endress + Hauser		Endress + Hauser	
			600 Kenrick, Suite B-3		600 Kenrick, Suite B-3	
			Houston, Texas 77060		Houston, Texas 77060	
			(713) 999-1991		(713) 999-1991	
Model			Cerabar S 430		Cerabar S 430	
Notes:						

Rev	Date	By	Ck'd	App'd	Description
A	8/23/96	EAH			Issued for Review
0	9/13/96	EAH			Issued for Bid

GSI Job No. G-1874
Issued: 9/13/96
Page 3 of 3

SPECIFICATION 3.10
INSTRUMENTATION SPECIFICATION:
PRESSURE TRANSMITTER

Experimental Controlled Release System (ECRS) II
Rice University, Houston, Texas

Tag No.			PIT-5		PIT-6	
Service:			Water		Water	
EFD No.			Figure 3.1		Figure 3.1	
Line No.			Groundwater Supply Pressure Sense		Groundwater Return Pressure Sense	
Process Conditions						
Flowrate (gpm)						
Average	Max		4	10	4	10
Pressure (psig)						
Average	Max		1	3	Atm	3
Temperature (°F)						
Average	Max		70	90	70	90
Vacuum Possibility			No		No	
Meter Specifications						
Function			Indicate & transmit pressure		Indicate & transmit pressure	
Display			Integral		Integral	
Materials of Construction						
Enclosure						
Process Seal Gasket			Viton		Viton	
Process Connection			Stainless steel		Stainless steel	
Sensor						
Diaphragm seal required?			No		No	
Conduit Entry						
Mounting			Surface, 1/2 in MNPT		Surface, 1/2 in MNPT	
Electrical						
Power Supply						
Grounding						
Electrical Classification			Not classified		Not classified	
Enclosure Classification						
Range			0 - 5 psig		0 - 5 psig	
Accuracy			Manufacturer standard		Manufacturer standard	
Output			4 - 20 mA		4 - 20 mA	
Display			LCD, units of in H_2O		LCD, units of in H_2O	
Purchasing						
Manufacturer			Endress + Hauser 600 Kenrick, Suite B-3 Houston, Texas 77060 (713) 999-1991		Endress + Hauser 600 Kenrick, Suite B-3 Houston, Texas 77060 (713) 999-1991	
Model			Cerabar S 430		Cerabar S 430	
Notes:						

Rev	Date	By	Ck'd	App'd	Description
A	8/23/96	EAH			Issued for Review
0	9/13/96	EAH			Issued for Bid

GSI Job No. G-1874
Issued: 9/13/96
Page 1 of 1

SPECIFICATION 3.11
INSTRUMENTATION SPECIFICATION:
ELECTROMAGNETIC FLOW METER

Experimental Controlled Release System (ECRS) II
Rice University, Houston, Texas

Tag No.		FIT-3			'
Service:		Water			
EFD No.		Figure 3.1			
Line No.		Groundwater Return from ECRS Tank			
Process Conditions					
Flowrate (gpm)					
Average	Max	6	10		
Pressure (psig)					
Average	Max	2	3		
Temperature (°F)					
Average	Max	70	100		
Fluid					
Sp. Gr.	Density (lb/ft³)	1			
Conductivity (µmhos/cm)		4,000 - 6,000			
Viscosity (cp)		1.002			
Vacuum Possibility		Yes			
Sensor / Transmitter Specifications					
Type		Electromagnetic			
Line Size (in)		1			
Materials of Construction					
Housing		316LSS / Epoxy Coated Aluminum			
Liner		Teflon			
Connections		316LSS			
Electrode		316LSS			
Connection					
Type		ANSI 150 Class Wafer or Flange			
Gasket		Viton Gasket			
Electronics Mounting		Self-contained			
Electrical					
Power Supply		85 - 260 VAC 50/60Hz			
Grounding					
Electrical Classification		Class 1, Division 2, Group C & D			
Enclosure Classification		NEMA 4X			
Output		4 - 20 mA			
Display		0 - 10 gpm			
Purchasing					
Manufacturer		Endress + Hauser			
		600 Kenrick, Suite B-3			
		Houston, Texas 77060			
		(713) 999-1991			
Model		Promag 33A T15HD1ED11D21A			
Notes:					

Rev	Date	By	Ck'd	App'd	Description
A	8/23/96	EAH			Released for Review
0	9/13/96	EAH			Released for Bid

GSI Job No. G-1874
Issued: 9/13/96
Page 1 of 2

SPECIFICATION 3.12
INSTRUMENTATION SPECIFICATION:
DIFFERENTIAL PRESSURE TRANSMITTER

Experimental Controlled Release System (ECRS) II
Rice University, Houston, Texas

Tag No.		FIT-1		FIT-2	
Service:		Sparge Air		Soil Vapor	
EFD No.		Figure 3.1		Figure 3.1	
Line No.		Sparge Mass Flow		SVE Mass Flow	
Process Conditions					
Flowrate (scfm)					
Average	Max	10	30	15	30
Meter Information					
Meter Diameter (in)		0.345		1.022	
Pipeline Diamter (in)		1.04		1.61	
Meter Taps		Flange		Flange	
Temperature (°F)					
Average	Max	80	—	80	—
Fluid					
Sp. Gr.	Density (lb/ft³)		0.075		0.075
Viscosity (cp)		0.018		0.018	
Vacuum Possibility		No		Yes	
Meter / Transmitter Specifications					
Function		Differential pressure		Differential pressure	
Application		Orifice plate flow measurement		Orifice plate flow measurement	
Range					
Output Characteristic		Square Root		Square Root	
Materials of Construction					
Wetted Materials		316 stainless steel		316 stainless steel	
Process Seal Gasket		Viton		Viton	
Fill Fluid		None		None	
Connections					
Process Connection		1/4 - 18 NPT oval flange		1/4 - 18 NPT oval flange	
Conduit Entry		1/2 in NPT		1/2 in NPT	
Mounting					
Type		2 in pipe		2 in pipe	
Electronics		Self-contained		Self-contained	
Electrical					
Power Supply		Loop Powered, 24 VDC		Loop Powered, 24 VDC	
Electrical Classification		Class 1, Division 2, Group C & D		Class 1, Division 2, Group C & D	
Output		4 - 20 mA		4 - 20 mA	
Display		Integral LCD		Integral LCD	
Purchasing					
Manufacturer		Endress + Hauser 600 Kenrick, Suite B-3 Houston, Texas 77060 (713) 999-1991		Endress + Hauser 600 Kenrick, Suite B-3 Houston, Texas 77060 (713) 999-1991	
Model		Deltabar S PMD 230		Deltabar S PMD 230	

Notes:
See attached orifice plate sizing calculations.

Rev	Date	By	Ck'd	App'd	Description
A	8/23/96	EAH			Release for Review
0	9/13/96	EAH	TLR		Release for Bid

GSI Job No. G-1874
Issued: 9/13/96
Page 2 of 2

SPECIFICATION 3.12
INSTRUMENTATION SPECIFICATION:
DIFFERENTIAL PRESSURE TRANSMITTER

Experimental Controlled Release System (ECRS) II
Rice University, Houston, Texas

Tag No.		LT-1			
Service:		Water			
EFD No.		Figure 3.1			
Line No.		Moisture Separator (V-2)			
Process Conditions					
Flowrate (gpm)					
Average	Max	4	10		
Pressure (in H$_2$O)					
Average	Max	15	60		
Temperature (°F)					
Average	Max	70	90		
Fluid					
Sp. Gr.	Density (lb/ft^3)	1			
Viscosity (cp)		1.002			
Vacuum Possibility		Yes			
Meter / Transmitter Specifications					
Function		Differential pressure			
Application		Level Measurement			
Range					
Output Characteristic		Linear			
Materials of Construction					
Wetted Materials		316 stainless steel			
Process Seal Gasket		Viton			
Fill Fluid		None			
Connections					
Process Connection		1/4 - 18 NPT oval flange			
Conduit Entry		1/2 in NPT			
Mounting					
Type		2 in pipe			
Electronics		Self-contained			
Electrical					
Power Supply		Loop Powered, 24 VDC			
Electrical Classification		Class 1, Division 2, Group C & D			
Output		4 - 20 mA			
Display		Integral LCD			
Purchasing					
Manufacturer		Endress + Hauser			
		600 Kenrick, Suite B-3			
		Houston, Texas 77060			
		(713) 999-1991			
Model		Deltabar S PMD 230			
Notes:					

Rev	Date	By	Ck'd	App'd	Description
A	8/23/96	EAH			Release for Review
0	9/13/96	EAH	TLR		Release for Bid

GSI Job No. G-1874
Issued: 9/13/96
Page 1 of 1

SPECIFICATION 3.13
INSTRUMENTATION SPECIFICATION:
TEMPERATURE SENSOR

Experimental Controlled Release System (ECRS) II
Rice University, Houston, Texas

GENERAL INFORMATION				
Complete Assembly	Yes		Element	
Head			Matl	Platinum
Screwed Cover	NA		Ice Pt Resistance	100 ohm
Explosion Proof	Yes		Temp Range	32 - 200 °F
Classification	Class 1, Div 2		Leads	Potted
Material	Stainless steel		Sheath Matl	1/4 in O.D. Stainless steel
Cond. Conn.	1/2 in		Mounting Thread	1/2 in MNPT
Nipple Size	1/2 in, close		Connection	3-wire lead wires

SPECIFIC REQUIREMENTS				
Tag No.	Process Conn.	Element Length	Single or Dual	Service
TE-1	1/2 MNPT	2 in	Single	Air Temperature (Spage)
TE-2	1/2 MNPT	2 in	Single	Air Temperature (SVE)
TE-3	1/2 MNPT	2 in	Single	Water Temperature (Groundwater Supply)
TE-4	1/2 MNPT	2 in	Single	Water Temperature (Groundwater Return)

Notes:
If minimum element length is greater than 2 in, use minimum length.

Rev	Date	By	Ck'd App'd	Description
A	8/23/96	TLR		Release for Review
0	9/13/96	EAH		Release for Bid

GSI Job No. G-1874
Issued: 9/13/96
Page 1 of 1

SPECIFICATION 3.14
INSTRUMENTATION SPECIFICATION:
LEVEL SWITCH

Experimental Controlled Release System (ECRS) II
Rice University, Houston, Texas

GENERAL	1	Tag No.		LSHH-1		LSHH-4	
	2	Service		Groundwater		Groundwater	
	3	Line No./Vessel No.		Moisture Separator (V-2)		Air Accumulator (V-1)	
BODY/CAGE	4	Body or Cage Matl		Stainless Steel		Stainless Steel	
	5	Conn Size & Location Upper		2000		2000	
	6	Type		1 in		1 in	
	7	Conn Size & Location Lower		MNPT		MNPT	
	8	Type		NA		NA	
	9	Case Mounting		NA		NA	
	10	Type		NA		NA	
	11	Rotatable Head		NA		NA	
	12	Orientation		Mfr std		Mfr std	
	13	Cooling Extension		NA		NA	
DISPLACER	14	Dimensions					
OR FLOAT	15	Insertion Depth		2 in		2 in	
	16	Displacer Extension		NA		NA	
	17	Disp . or Float Material		Poly Pro		Poly Pro	
	18	Displacer Spring/Tube Matl					
XMTR/CONT.	19	Function		Switch		Switch	
	20	Output		Contact Closure		Contact Closure	
	21	Control Modes		NA		NA	
	22	Differntial		NA		NA	
	23	Outpu Action: Level Rise		NA		NA	
	24	Mounting		NA		NA	
	25	Enclosure Class		NA		NA	
	26	Elec. power or Air Supply		208 V , 1Ø		208 V , 1Ø	
SERVICE	27	Upper Liquid					
	28	Lower Liquid		NA		NA	
	29	sp. gr.: Upper	Lower	NA	1	NA	1
	30	Press. Max.	Normal	NA	5 psi	NA	5 psi
	31	Temp. Max.	Normal	NA	90°F	NA	90°F
OPTIONS	32	Airset	Supply Gauge	NA	NA	NA	NA
	33	Gauge Glass Connections		NA		NA	
	34	Gauge Glass Model No.		NA		NA	
	35	Contacts : No.	Form	1	C	1	C
	36	Contact Rising		5 A, 250 VAC		5 A, 250 VAC	
	37	Action of Contacts		SPDT		SPDT	
	38	Manufacturer		W.E. Anderson		W.E. Anderson	
	39	Model No.		L6EPB-S-S-3-0		L6EPB-S-S-3-0	

NOTES
1. NA = Not applicable.

GSI Job No. G-1874
Issued: 9/13/96
Page 1 of 1

SPECIFICATION 3.15
INSTRUMENTATION SPECIFICATION:
PRESSURE GAUGE

Experimental Controlled Release System (ECRS) II
Rice University, Houston, Texas

GENERAL INFORMATION				
Type	Direct Reading	Socket Material	316 Stainless Steel	
Mounting	Flush Panel/Local (See Fig 3.1)	Connection	1/2 in - 14 NPT	
Dial Diamter (in)	4.5	Movement	Stainless Steel	
Case Material	Cast Aluminum	Diaphragm Seal	None	
Ring Type	Steel	Purchasing		
Blowout Protection	None	Manufacturer	US Gauge Divison	
Lens	High-Temperature Acrylic		Sellersville, PA 18960	
Nominal Accuracy Req'd	± 0.5% of Span (Grade 2A)		(215) 257-6531	
Pressure Element	Bourdon Tube			
Element Material	316 Stainless Steel	Model	1933 4-1/2 in Solfrunt	

SPECIFIC REQUIREMENTS				
Quantity	Tag No.	Range	Operating Pressure (psia)	Service
1	PG-1	0 - 60 in H_2O	18	Groundwater Gradient Control Inlet
1	PG-2	0 - 60 in H_2O	18	Groundwater Gradient Contol Outlet
1	PG-3	0 - 100 psig	75	Groundwater Pump Discharge
1	PG-4	0 - 100 psig	75	Groundwater Filter Discharge
1	PG-5	0 - 15 psig	20	Groundwater Treatment Supply
1	PG-6	0 - 5 psi	20	Sparge Supply
1	PG-7	0 - 30 in Hg (vac)	15	Soil Vapor Extraction
1	PG-8	0 - 200 psig	165	Main Air Supply
1	PG-9	0 - 100 psig	115	Utility Air Supply
1	PG-10	0 - 15 psig	15	Groundwater Treatment Discharge
1	PG-11	0 - 15 psig	30	Utility Pump

Notes:

Rev	Date	By	Ck'd	App'd	Description
0	9/13/96	EAH	TLR		Release for Bid

GSI Job No. G-1874
Issued: 9/13/96
Page 1 of 1

SPECIFICATION 3.16
INSTRUMENTATION SPECIFICATION:
FLOW GAUGE

Experimental Controlled Release System (ECRS) II
Rice University, Houston, Texas

Tag No.			FG-1			
Service:			Water (see notes)			
EFD No.			Figure 3.1			
Line No.			Groundwater Supply to ECRS Tank			
Process Conditions						
Flowrate (gpm)						
Average	Maximum		2	10		
Pressure (psig)						
Average	Maximum		10	40		
Temperature (°F)						
Average	Maximum		70	100		
Fluid						
Sp. Gr.	Density (lb/ft³)		1			
Conductivity (μmhos/cm)			>5			
Viscosity (cp)			1.002			
Vacuum Possibility			No			
Meter / Transmitter Specifications						
Type			Float and Tapered Tube			
Line Size (in)			1			
Materials of Construction						
Wetted Parts			317 Stainless Steel			
O-Rings			Viton®			
Spring			316 Stainless Steel			
Connection Type			3/4 in NPT			
Display			Analog , 160° sweep, 0 -10 gpm			
Range			0 - 8 gpm			
Accuracy			±3% accuracy, ±1 repeatability			
Purchasing						
Manufacturer			King Instrument Co. c/o Controls Warehouse 356 Cypress Road Ocala, FL 34472 (201) 992-1400			
Model			7712-230-718 (10 gpm)			

Notes:
Process water contains dilute (<1%) concentrations of benzene, toluene, ethylbenzene, and xylenes.

Rev	Date	By	Ch'd	App'd	Description
A	8/23/96	EAH			Release for Review
0	9/13/96	EAH			Release for Bid

GSI Job No. 1874
Issued: 9/13/96
Page 1 of 4

SPECIFICATION 3.17
ELECTRICAL SPECIFICATIONS

Enclosure

The electrical enclosure shall be a free-standing NEMA 4 steel enclosure with a gray polyester coating over phosphatized surfaces inside and out. Nominal overall dimensions of the enclosure shall be 72-in tall, 31-in wide, and 24-in deep. The enclosure shall be provided with lifting eyes and mounting hardware for securing to the process equipment skid. The enclosure shall be Hoffman model A-72H3124FS.

Swing-out Rack Mounting Frame

A 19-in. wide swing-out rack mounting frame shall be provided for the above enclosure. The rack mounting frame shall be Hoffman model A-7230SOF19.

Rack Panels

19-in wide steel rack panels shall be provided as necessary to accomodate the electrical components to be mounted on the rack frame. The panels shall conform to standard E.I.A. dimensions. Rack panels shall have a gray finish. All hardware required to mount the panels shall be provided by the contractor. Rack panels shall be Hoffman model A-19RP*U. (Substitute the number of 1.75-in high units.)

Enclosure Panels

Side and rear panels shall be provided as necessary to accomodate the electrical components to be mounted within the enclosure. The panels shall be Hoffman model A-72SMP14, A-72SMP20, A-72P30F1, or A-72P30F2 as appropriate.

Power Transformer

A three-phase 7.5 kVA power transformer shall be provided and mounted within the electrical enclosure. The transformer shall be 480 V delta by 208/120V wye connected. Overcurrent protection shall be in accordance with manufacturer's reccommendations but not smaller than that shown on Figure 3.3, Electrical One-line Diagram.

GSI Job No. 1874
Issued: 9/13/96
Page 2 of 4

SPECIFICATION 3.17
ELECTRICAL SPECIFICATIONS

Constant Voltage Transformer

A single-phase 1 kVA constant voltage transformer shall be provided and mounted within the electrical enclosure. Overcurrent protection shall be in accordance with manufacturer's reccommendations but not smaller than that shown on Figure 3.3, Electrical One-line Diagram. The manufacturer of the constant voltage transformer shall be SOLA.

Linear Power Supply

A 24 volt, 10 amp, linear power supply shall be provided to supply loop powered instrumentation. The power supply shall be rack mounted with voltmeter, ammeter, and fuse accesible from the front of the rack mounted panel. Linear power supply shall be Acopian model V24PT10AF.

Combination Motor Starters

Combination motor starters consisting of disconnect switch, motor contactor, and motor overload device shall be provided for Groundwater Pump, P-1, and Utility Pump, P-2. These combination motor starters shall be horsepower rated and sized according to the nameplate information on the respective motors for P-1 and P-2.

Terminal Blocks

Terminal blocks shall be provided for power distribution and instrumentation tie-ins. Terminal Blocks shall be WAGO, Allen-Bradley, or equivalent.

Instrumentation Fuses and Fuse Holders

Instrumentation fuses and fuse holders shall be provided as shown on Figures 3.3 and 3.4, Electrical One-line and Ladder Diagram, repsectively. Fuses shall be rack panel mounted in individual fuse holders or they may be integral to terminal blocks. Fuses shall be G-type cartridge fuses, Bussman or equivalent.

Process Equipment Disconnect Switch

A 3-pole, 45 amp, unfused disconnect switch, suitable for 480 volt service, shall be provided as shown on Figure 3.3, Electrical One-line Diagram. The

GSI Job No. 1874
Issued: 9/13/96
Page 3 of 4

SPECIFICATION 3.17
ELECTRICAL SPECIFICATIONS

switch shall be mounted as shown on Figure 3.6, General Arrangement Process Equipment Skid: Left Front Isometric.

Instrumentation Building Disconnect Switch

A 3-pole, 15 amp, fused disconnect switch, suitable for 480volt service, shall be provided as shown on Figure 3.3, Electrical One-line Diagram. The switch shall be mounted as shown on Figure 3.6, General Arrangement Process Equipment Skid: Left Front Isometric.

Utility Outlets

Two duplex GFCI utility outlets shall be provided and mounted on the Process equipment skid where convenient at a height of at least 18-in above the top of steel on the Process Equipment Skid. Outlets shall be enclosed in weatherproof enclosures.

Overcurrent Protection

Overcurrent protection devices called out as circuit breakers on Figure 3.3, Electrical One-line Diagram, shall have integral magnetic or thermomagnetic trip devices. Only GFCI circuit breakers shall be used.

Sparge Timer

A single preset digital timer shall be provided and installed in the door of the enclosure. The timer shall operate in a timed output at zero, automatic reset to preset after timed output mode. The timer face shall maintain the NEMA 4 rating of the enclosure. The timer shall be able to accomodate intervals ranging from 0 seconds to 99 hours and 59 minutes for the preset countdown timer and the timed output. Timer output shall be via a Form C relay rated at 5A @ 120/240 VAC. Run input (enable) shall be via contact closure. the timer shall be Red Lion model LIBT1E00.

Instrument Access Door

A single instrument access door shall be provided and installed in the enclosure door to allow access to the Data recorder, UIR-1, without opening the main enclosure door. The instrument door shall maintain the NEMA 4 rating of the enclosure. The instrument access door shall be BEBCO model PIAD sized to allow complete access to the front of the data recorder.

GSI Job No. 1874
Issued: 9/13/96
Page 4 of 4

SPECIFICATION 3.17
ELECTRICAL SPECIFICATIONS

Panel Operators

Panel operators (HOA switches, Emergency stop pushbutton, etc.) shall maintain the NEMA 4 rating of the enclosure. Panel operators shall be Allen-Bradley with sealed contact blocks.

Panel Cooling

The electrical panel shall be cooled with a vortex enclosure cooler. The vortex cooler shall be ITW VORTEC sized to provide appropriate cooling.

Miscellaneous Items

Miscellaneous items such as wire, mounting hardware, cable glands, etc., shall be supplied as necessary to complete the electrical installation. All items shall conform to applicable local codes and the NEC.

GSI Job No. G-1874
Issued: 9/13/96
Page 1 of 1

SPECIFICATION 3.18
INSTRUMENTATION SPECIFICATION
FLOW SWITCH (FSLL-1)

An automatic explosion-proof flow switch shall be vane operated to actuate a single pole double throw snap switch. Motion of the vane shall actuate the switch by the action of a magnet which controls the switch inside of the one-piece sealed switch body. Switches shall be W.E. Anderson Catalog No. V6 Standard with specified materials and pipe size.

Actuation Flow:	3.0 GPM
Deactuation Flow:	1.75 GPM
Temperature Limits:	220°F
Operating Pressure:	100 PSI
Piping Connection:	1-in Female NPT
Electrical Rating:	One SPDT snap action micro switch. 5A @ 125/250 VAC (resistive) 0.5A @ 125 VDC (resistive)
Wiring Connections:	3 each 18-in leads, color coded. Normally open, normally closed, and common.
Switch Body:	Lower Housing - Stainless steel Upper Housing - Brass
Vane:	Stainless steel, standard 11/16-in wide x .020
Installation:	Index arrow in direction of flow.
Mounting:	Any position.

GSI Job No. G-1874
Issued: 9/13/96
Page 1 of 1

SPECIFICATION 3.19
INSTRUMENTATION SPECIFICATION:
PRESSURE SWITCH (PSHH-4)

A pressure switch shall be diaphragm operated with fully adjustable setpoint. It shall be weather-proof and explosion-proof. Switch shall have visible set point indicator protected by clear polycarbonate window. Switch shall have factory installed explosion-proof seal with 18-in leads. Unit shall be Mercoid Corp. Model 1003E-B2-J.

Temperature Limits:	170°F
Pressure Connection:	1/4-in
Electrical Rating:	1/8 HP, 12A @ 125 VAC (resistive) 1/4 HP, 12A @ 250 VAC (resistive) 0.5A @ 125 VDC (resistive)
Wiring Connections:	3 each 18-in leads, color coded. Normally open, normally closed, and common.
Set Point Adjustment:	Thumbwheel type inside housing - isolated from electrical chamber - visible scale.
Housing:	Die-cast, copper-free aluminum. Explosion-proof and weather-proof.
Diaphragm:	Stainless steel
Calibration Spring:	Plated steel.
Mounting:	4 holes for 1/4-in bolts.
Installation:	Any position.

GSI Job No. G-1874
Issued: 9/13/96
Page 1 of 1

SPECIFICATION 3.20
INSTRUMENTATION SPECIFICATION
SIGHT GLASS (SG-1)

Sight glass shall be a flow-through type sight flow indicator. Sight glass shall allow observation of flowing stream from either side of pipe. Unit shall be Ernst Gage Co. Figure E-57-0 stainless steel, viton gaskets, tempered glass, with 1 1/2 -in process connections.

Temperature Limits: 170°F

Process Connection: 1 1/2-in

Body: Stainless steel.

Glass: Tempered glass.

Gasket: Viton

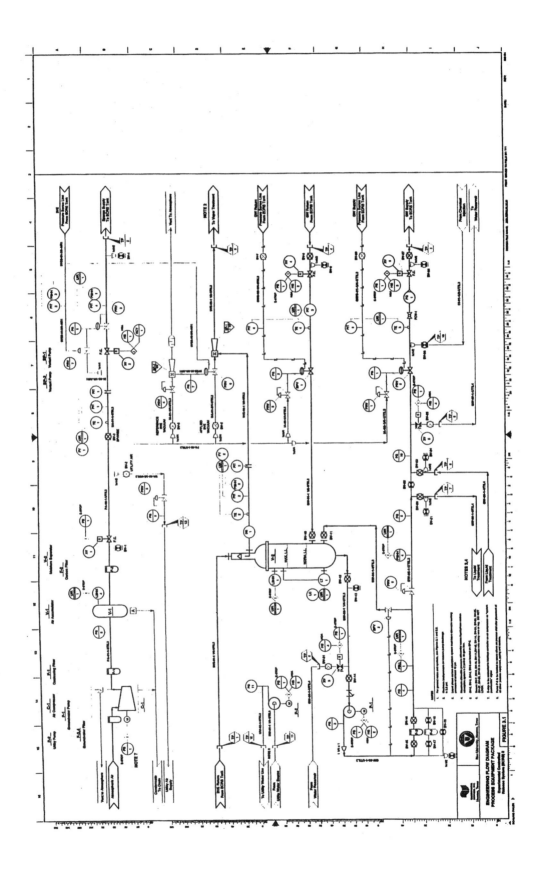

FIGURE 3.1

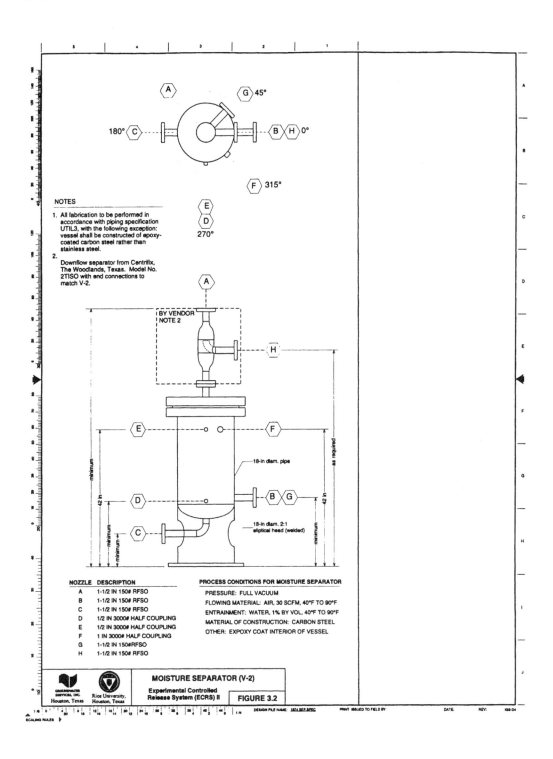

NOTES

1. All fabrication to be performed in accordance with piping specification UTIL3, with the following exception: vessel shall be constructed of epoxy-coated carbon steel rather than stainless steel.

2. Downflow separator from Centrifix, The Woodlands, Texas. Model No. 2TISO with end connections to match V-2.

NOZZLE	DESCRIPTION
A	1-1/2 IN 150# RFSO
B	1-1/2 IN 150# RFSO
C	1-1/2 IN 150# RFSO
D	1/2 IN 3000# HALF COUPLING
E	1/2 IN 3000# HALF COUPLING
F	1 IN 3000# HALF COUPLING
G	1-1/2 IN 150#RFSO
H	1-1/2 IN 150# RFSO

PROCESS CONDITIONS FOR MOISTURE SEPARATOR

PRESSURE: FULL VACUUM

FLOWING MATERIAL: AIR, 30 SCFM, 40°F TO 90°F

ENTRAINMENT: WATER, 1% BY VOL, 40°F TO 90°F

MATERIAL OF CONSTRUCTION: CARBON STEEL

OTHER: EXPOXY COAT INTERIOR OF VESSEL

GROUNDWATER SERVICES, INC. Houston, Texas

Rice University, Houston, Texas

MOISTURE SEPARATOR (V-2)

Experimental Controlled Release System (ECRS) II

FIGURE 3.2

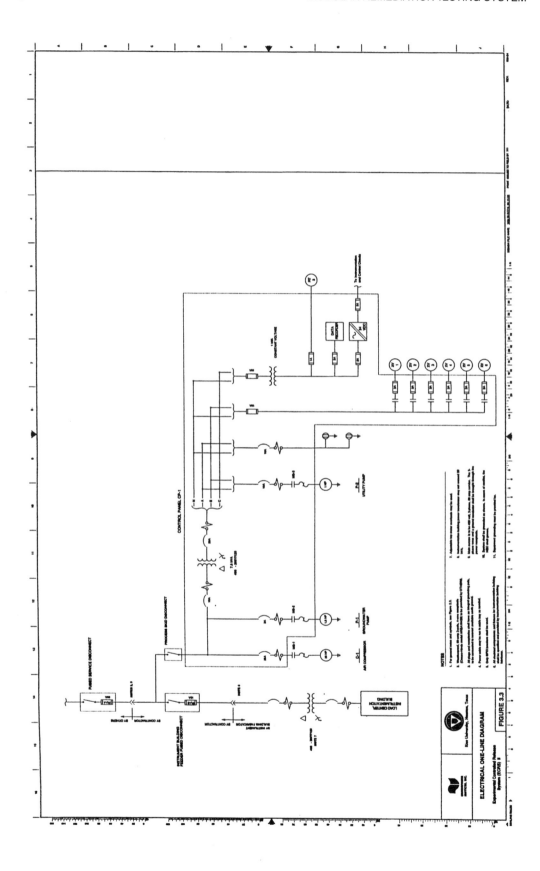

ELECTRICAL ONE-LINE DIAGRAM

Experimental Controlled Release System (ECRS) II

FIGURE 3.3

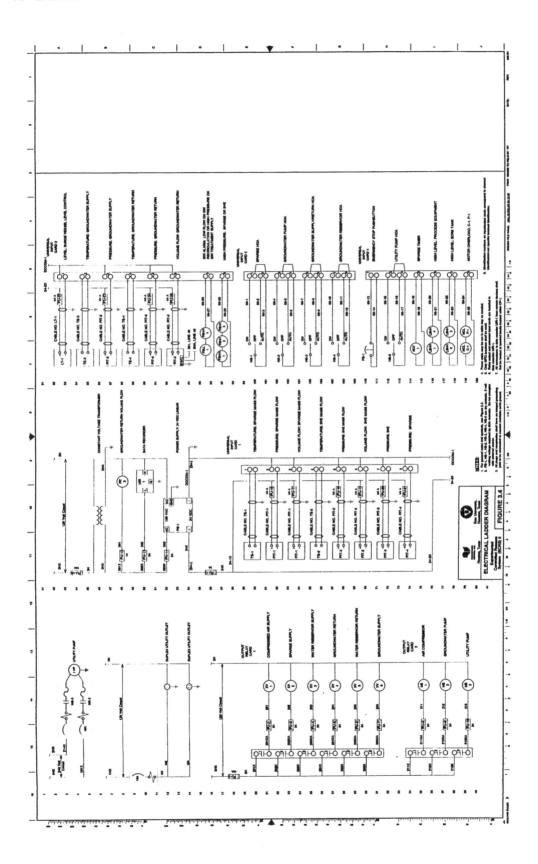

ELECTRICAL LADDER DIAGRAM

FIGURE 3.4

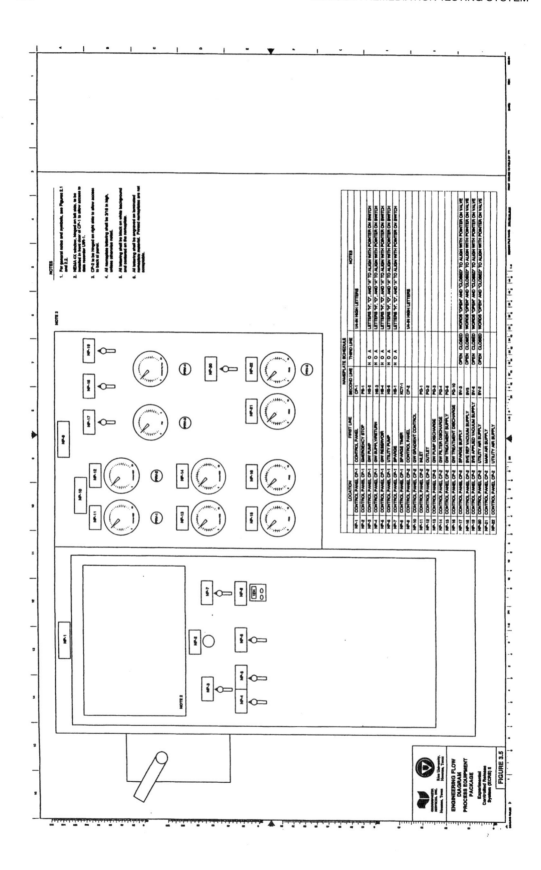

FIGURE 3.5

ENGINEERING FLOW DIAGRAM
PROCESS EQUIPMENT PACKAGE

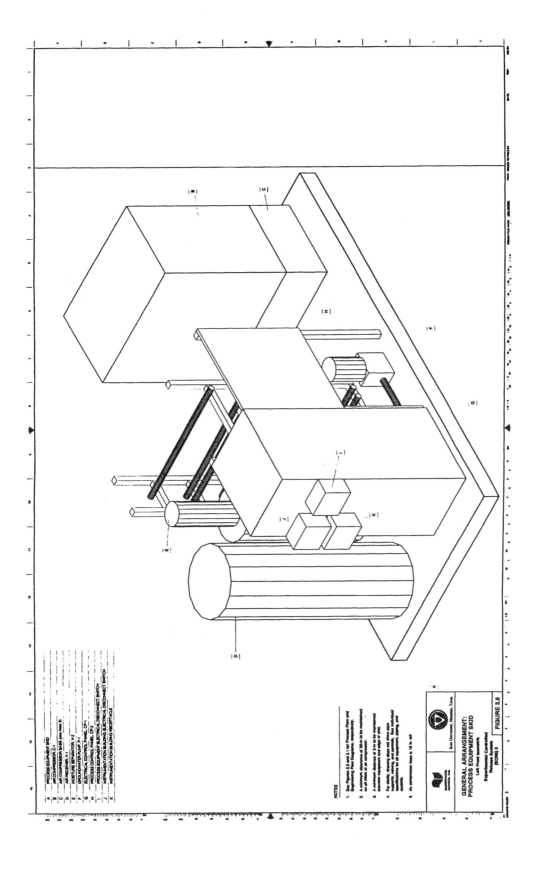

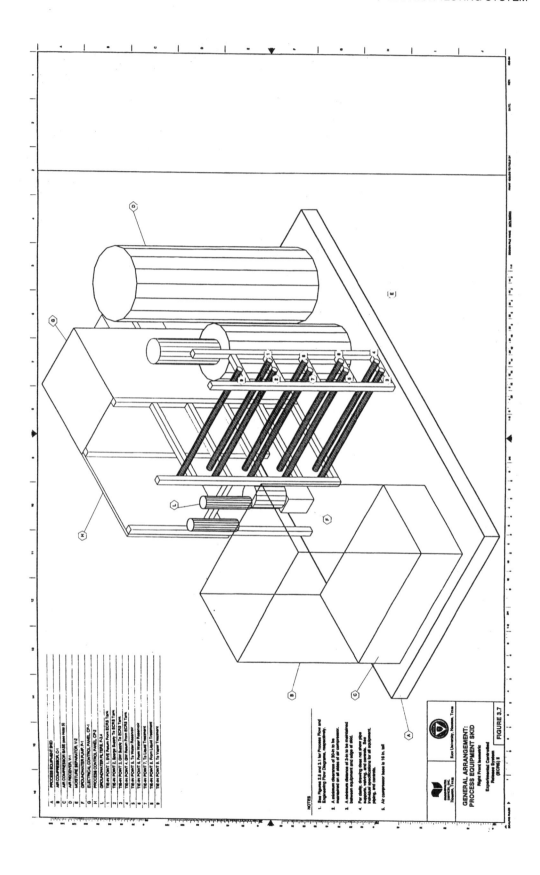

GSI Job No. G-1874
September 13, 1996

ENGINEERING SPECIFICATIONS

Experimental Controlled Release System (ECRS) II
DOD-AATDF/Rice University, Houston, Texas

SECTION 4.0 SPECIFICATIONS

Specification 4.1   Level Switch
Specification 4.2   Pressure Relief Valve

SECTION 4.0 FIGURES

Figure 4.1   Engineering Flow Diagram: ECRS Tank
Figure 4.2   General Arrangement: ECRS Tank
Figure 4.3   ECRS Tank Details
Figure 4.4   Top Details

GSI Job No. G-1874
September 13, 1996

4.0   ECRS TANK

The ECRS tank (T-1) will be fabricated by modifying a 27 cubic yard rectangular steel sludge tank. An engineering flow diagram for the ECRS tank is provided on Figure 4.1. As shown on Figures 4.2 and 4.3, the tank will be fitted with flanged and tubing nozzles and equipped with valves, tee-strainers, and sight gauges. At the time of installation, the ECRS tank will be provided with a flexible top to be provided and installed by others. In order to complete this installation, the contractor will fabricate and supply the clamps and nozzles described on Figure 4.4. Instrumentation requirements for level switches and a pressure relief valve for the ECRS tank are provided on Specifications 4.1 and 4.2.

GSI Job No. G-1874
Issued: 9/13/96
Page 1 of 1

SPECIFICATION 4.1
INSTRUMENTATION SPECIFICATION:
LEVEL SWITCH

Experimental Controlled Release System (ECRS) II
Rice University, Houston, Texas

GENERAL	1	Tag No.		LSHH-2		LSHH-3	
	2	Service		Groundwater		Groundwater	
	3	Line No./Vessel No.		ECRS Tank (T-1)		ECRS Tank (T-1)	
BODY/CAGE	4	Body or Cage Matl		Stainless Steel		Stainless Steel	
	5	Conn Size & Location Upper		2000		2000	
	6	Type		1 in		1 in	
	7	Conn Size & Location Lower		MNPT		MNPT	
	8	Type		NA		NA	
	9	Case Mounting		NA		NA	
	10	Type		NA		NA	
	11	Rotatable Head		NA		NA	
	12	Orientation		Mfr std		Mfr std	
	13	Cooling Extension		NA		NA	
DISPLACER OR FLOAT	14	Dimensions					
	15	Insertion Depth		2 in		2 in	
	16	Displacer Extension		NA		NA	
	17	Disp . or Float Material		Poly Pro		Poly Pro	
	18	Displacer Spring/Tube Matl					
XMTR/CONT.	19	Function		Switch		Switch	
	20	Output		Contact Closure		Contact Closure	
	21	Control Modes		NA		NA	
	22	Differntial		NA		NA	
	23	Outpu Action: Level Rise		NA		NA	
	24	Mounting		NA		NA	
	25	Enclosure Class		NA		NA	
	26	Elec. power or Air Supply		208 V , 1Ø		208 V , 1Ø	
SERVICE	27	Upper Liquid					
	28	Lower Liquid		NA		NA	
	29	sp. gr.: Upper	Lower	NA	1	NA	1
	30	Press. Max.	Normal	NA	5 psi	NA	5 psi
	31	Temp. Max.	Normal	NA	90°F	NA	90°F
OPTIONS	32	Airset	Supply Gauge	NA	NA	NA	NA
	33	Gauge Glass Connections		NA		NA	
	34	Gauge Glass Model No.		NA		NA	
	35	Contacts : No.	Form	1	C	1	C
	36	Contact Rising		5 A, 250 VAC		5 A, 250 VAC	
	37	Action of Contacts		SPDT		SPDT	
	38	Manufacturer		W.E. Anderson		W.E. Anderson	
	39	Model No.		L6EPB-S-S-3-0		L6EPB-S-S-3-0	

NOTES
1. NA = Not applicable.

GSI Job No. 1874
Issued: 9/13/96
Page 1 of 1

SPECIFICATION 4.2
PRESSURE SAFETY VALVE, PSV-1

A pressure safety valve shall be provided to prevent the inflation and rupture of the flexible top of the ECRS Tank, T-1. The valve shall be vented to atmosphere and shall be suitable for "all-weather" operation. The process connection shall be a 2-in 150# class flange. The materials of construction shall be carbon steel body with stainless steel trim. The valve seat insert shall be Viton. The relieving perssure shall be 1 oz/sq. in. The pressure safety valve shall be a Whesoe Varec model 2011-2-3-V-RF-0-02-VB.

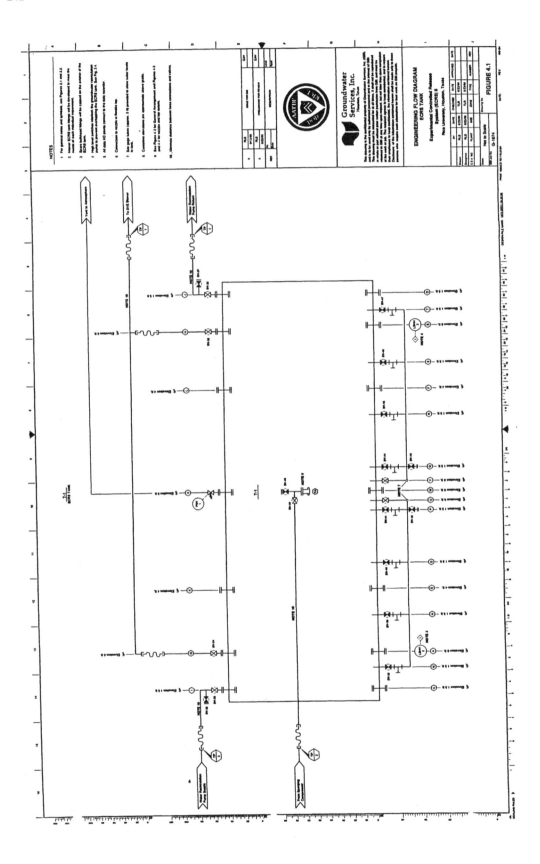

FIGURE 4.1

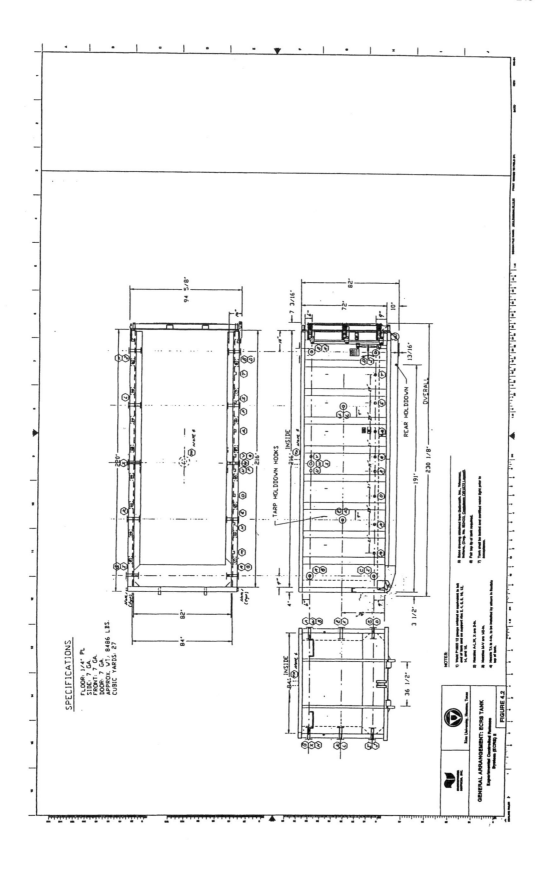

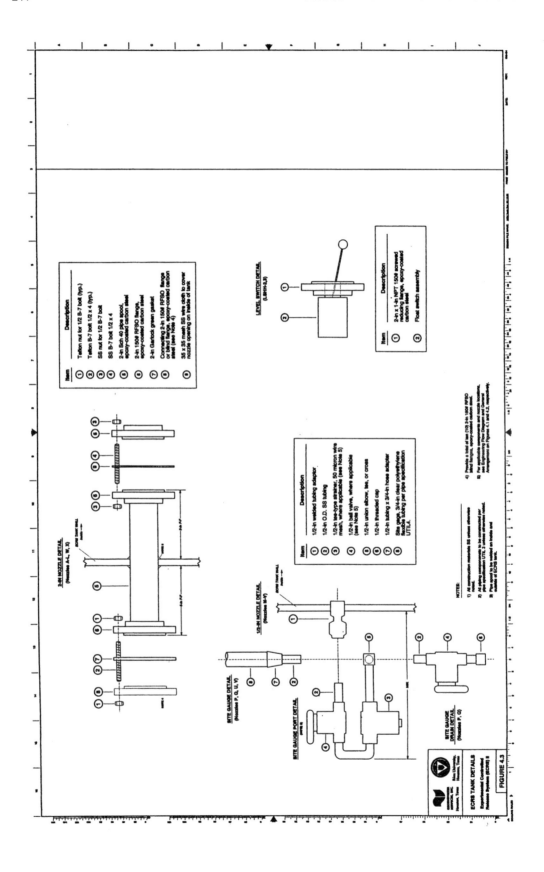

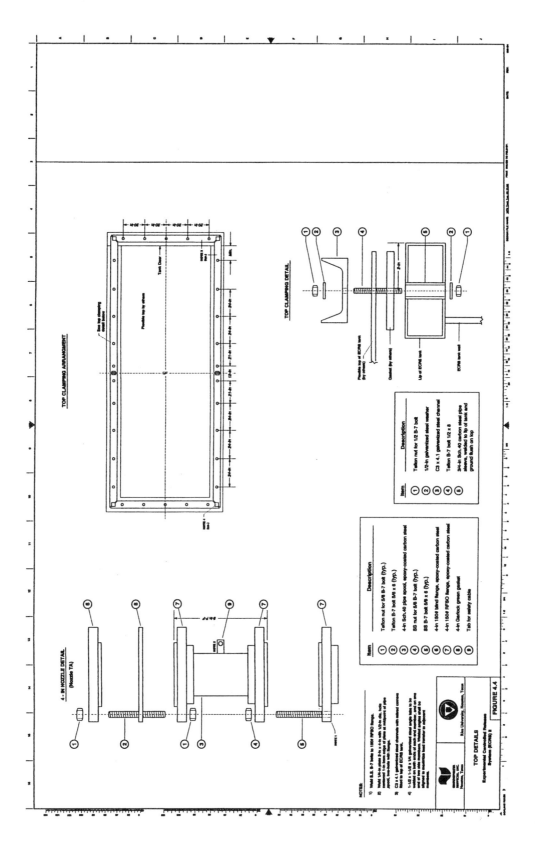

GSI Job No. G-1874
September 13, 1996

5.0   ANCILLARY COMPONENTS

Additional equipment required to complete the system is shown in the engineering flow diagram on Figure 5.1 and will comprise the following: 1) a water storage tank, 2) a chemical mixing tank assembly, and 3) air and water treatment units. An engineering flow diagram for all ancillary equipment is included as Figure 5.1. A 1,500 gal polyethylene tank (T-2) will be supplied and provided with valves and flexible hose connections as described on Specification 5.2 and Figure 5.2. The contractor will also provide a 50-gal tank (T-3) with mixer for preparing chemical solutions (see Specification 5.2 and Figure 5.2). Air and water extracted from the ECRS tank will likely require treatment to remove contaminants prior to discharge into the atmosphere or release into an appropriate wastewater collection system.

GSI Job No. G-1874
September 13, 1996

ENGINEERING SPECIFICATIONS

Experimental Controlled Release System (ECRS) II
DOD-AATDF/Rice University, Houston, Texas

SECTION 5.0 SPECIFICATIONS

GSI Job No. G-1874
Issued: 9/13/96
Page 1 of 1

SPECIFICATION 5.1
EQUIPMENT SPECIFICATION: PUMPS

Experimental Controlled Release System (ECRS) II
Rice University, Houston, Texas

Tag No.				P-3				
Service:				Various aqueous chemical solutions				
EFD No.				Figure 5.1				
Line No.				Chemical mixing tank (T-3) discharge				
Process Conditions								
Flowrate (gpm)								
Max	Norm	Min		1.2E-01	3.5E-02			
Temperature (°F)								
Max	Norm	Min		100	70	40		
Fluid								
Sp. Gr.				<1.5				
Viscosity				<100 cp				
Duty Cycle				Continuous				
Service (%)				100				
Moisture (%)				100				
Abrasives				None				
Solids								
Fraction (%)				<1				
Size (μm)				<1				
Unit Specifications								
Type				Positive Displacement Metering				
Dimensions (L x W x H)				10.75 in x 5.72 in x 8 in				
Materials of Construction				316 SS				
Gaskets/Seals				Teflon				
Connections								
Inlet (in)				0.25				
Outlet (in)				0.25				
Discharge Pressure (ft H_2O)								
Max	Norm	Min		5	3	2		
Power Requirements				115 V ac				
Purchasing								
Manufacturer				LMI / Milton Roy Acton, MA 01720 (508) 264-9172				
Model				B141-217				
Notes:								

Rev	Date	By	Chk'd	App'd	Description
A	8/23/96	RLB	EAH	Preliminary for review	
0	9/13/96	RLB	EAH	Issue for bid	

GSI Job No. G-1874
Issued: 8/23/96
Page 1 of 2

SPECIFICATION 5.2
EQUIPMENT SPECIFICATION: TANKS

Experimental Controlled Release System (ECRS) II
Rice University, Houston, Texas

Tag No.	T-2	T-3
Service:	Water	Water
EFD No.	Figure 5.1	Figure 5.1
Tank Specifications:		
Function	Clean process water storage	Chemical batch mixing tank w/ stand
Type	Vertical cylindrical	Open top, cone bottom, cylindrical
Capacity (minimum)	1500 gal.	50 gal.
Dimensions:		
Diameter (maximum)	72 in.	24 in.
Height (maximum)	132 in.	42 in.
Materials of Construction	XLPE (Sp. Grav. = 1.5)	XLPE (Sp. Grav. = 1.5)
Color	Opaque (blue preferred)	Translucent white/clear
Through-wall Fitting Specifications:		
Type	Flanged	Welded half coupling
Size	2 in. (except for site gauge, 3/4 in.)	1/2 in.
Materials of Construction:	SS	PE
Encapsulated Bolt/Capsule	SS/Viton	
Gaskets	Viton	
Siphon Drain	PVC	
Accesory Specifications:		
Siphon Drain		Not Applicable
Size	2 in.	
Materials of Construction:	PVC Sch 40	
Fill Line Assembly		Not Applicable
Type	Through top of tank with union	
Size	2 in.	
Materials of Construction:	PVC Sch 40	
Bracket Bolt/Capsule	SS/Viton	
Ball Valve Seals	Viton	
U-Vent Assembly:		Not Applicable
Type	180° U-Bend	
Size	2 in.	
Materials of Construction:	PVC Sch 40	
Sight Gauge Assembly:		Not Applicable
Type	Double Valve	
Size	3/4 in.	
Materials of Construction:	Clear flexible PVC	
Ball Valve Seals	Viton	
Vertical Tie-downs:		Not Applicable
Type	4-in bands with 2 x 20-in legs	
Materials of Construction:		
Bands/Legs	Epoxy-painted mild steel	
Bolts	SS	
Mixer:	Not Applicable	
Motor		1/4 hp, 115/230 V, TEFC
Materials of Construction		SS
Shaft		1/2-in dia. x 30-in
Propeller		4-in dia.

GSI Job No. G-1874
Issued: 8/23/96
Page 2 of 2

SPECIFICATION 5.2
EQUIPMENT SPECIFICATION: TANKS

Experimental Controlled Release System (ECRS) II
Rice University, Houston, Texas

Tag No.	T-2	T-3
Purchasing:		
Manufacturer:	Assman Corp. of America 300 North Taylor Road Garrett, IN 46738 (219) 357-3181	Chem-Tainer Industries, Inc. 361 Neptune Ave. West Babylon, NY 11704 (516) 661-8300
Model Numbers:		
Tank	ICT 1500	TC2236CA
Fitting		FB series
Mixer		CTBN 10
Stand w/ Mixer Mount Bar		TC2236CM (see Note 2)

Notes:
1) XLPE = Crosslinked Polyethylene; SS = Stainess Steel; PVC = Polyvinyl Chloride
2) Locking casters to support 1000 lb. load shall be supplied and attached to mixing tank stand by contractor.

Rev	Date	By	Chk'd	App'd	Description
A	8/23/96	RLB	EAH		Preliminary for review
0	9/13/96	RLB	EAH		Issue for bid

GSI Job No. G-1874
September 13, 1996

ENGINEERING SPECIFICATIONS

Experimental Controlled Release System (ECRS) II
DOD-AATDF/Rice University, Houston, Texas

SECTION 5.0 FIGURES

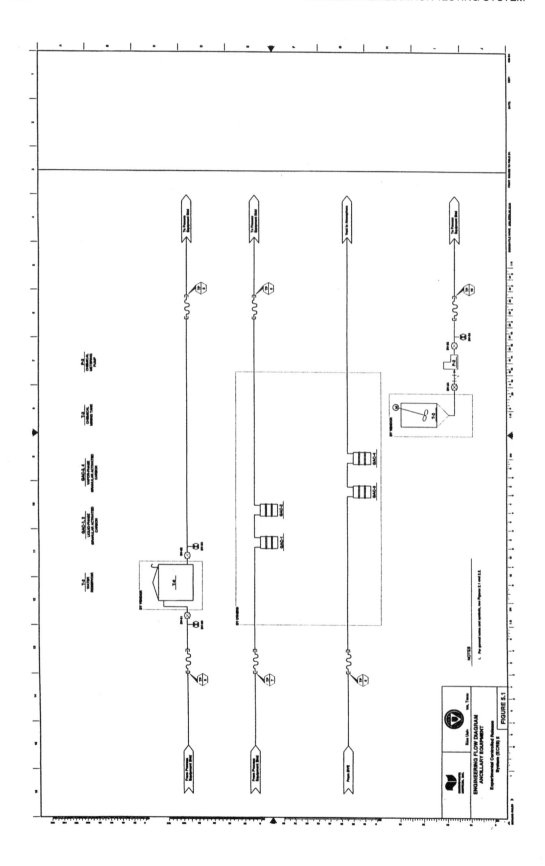

FIGURE 5.1

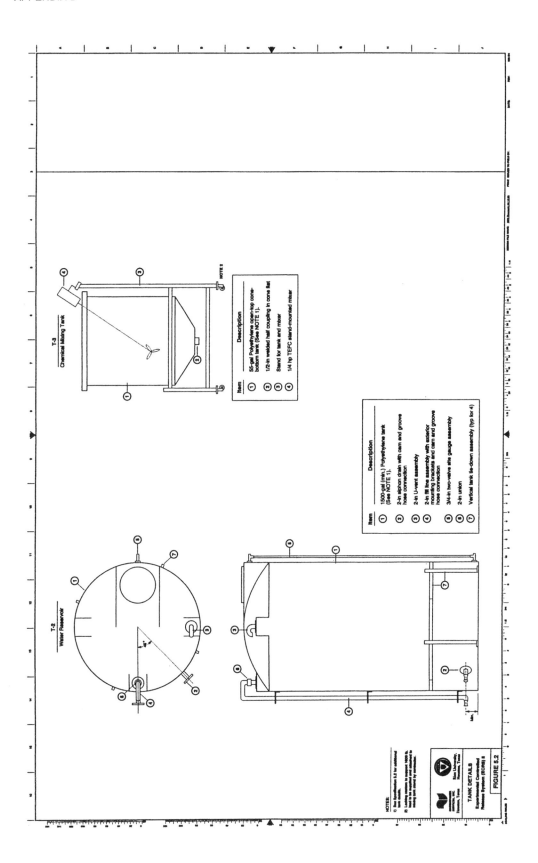

Appendix E

Final ECRS Unit 2 Engineering Diagrams

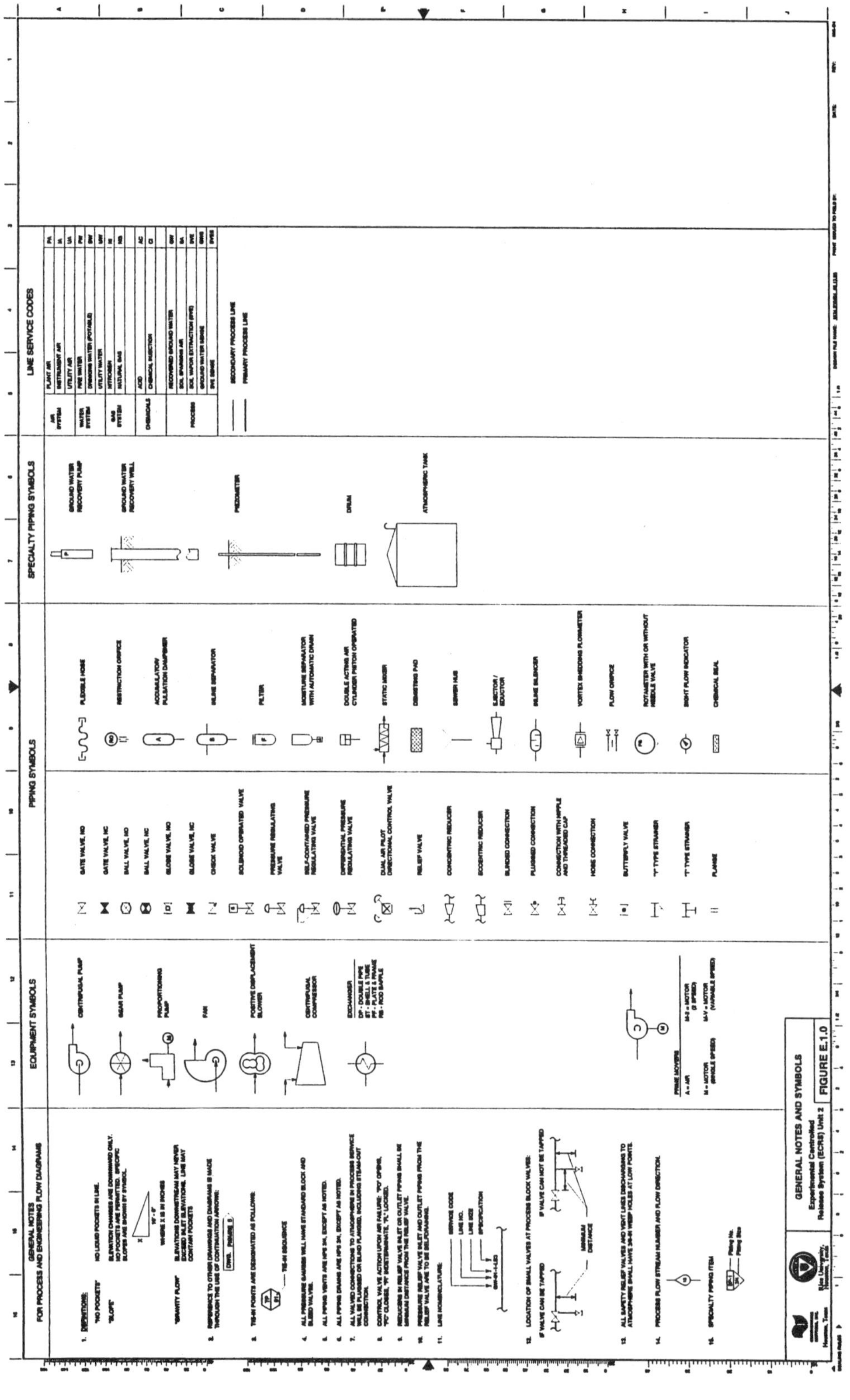

Figure E1.0 General notes and symbols.

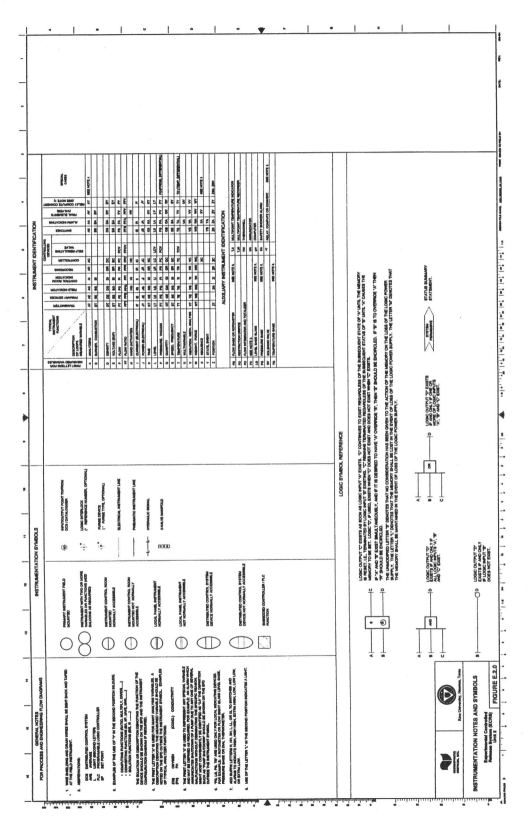

Figure E2.0   Instrumentation notes and symbols.

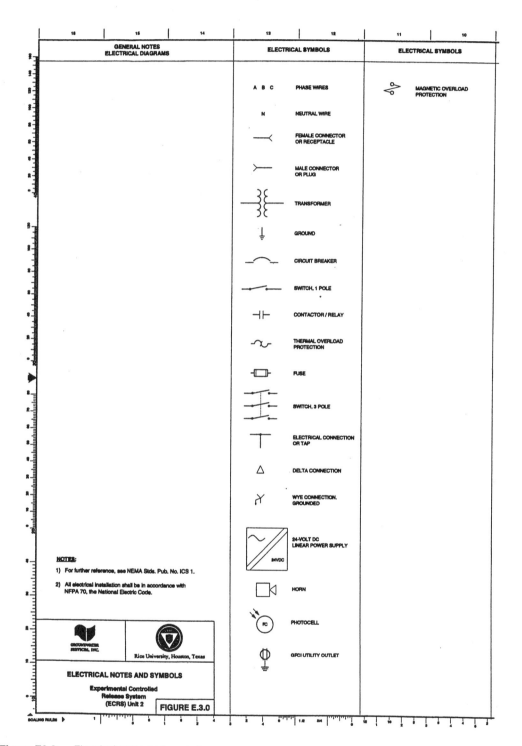

Figure E3.0   Electrical notes and symbols.

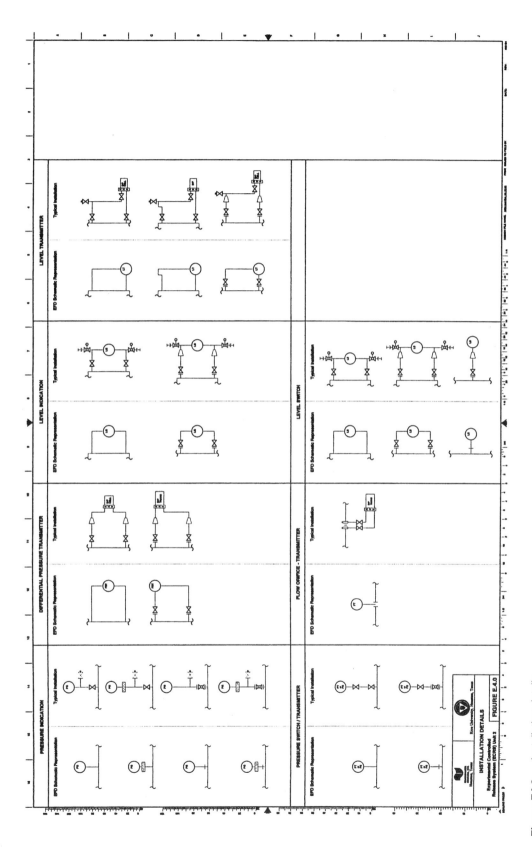

Figure E4.0 Installation details.

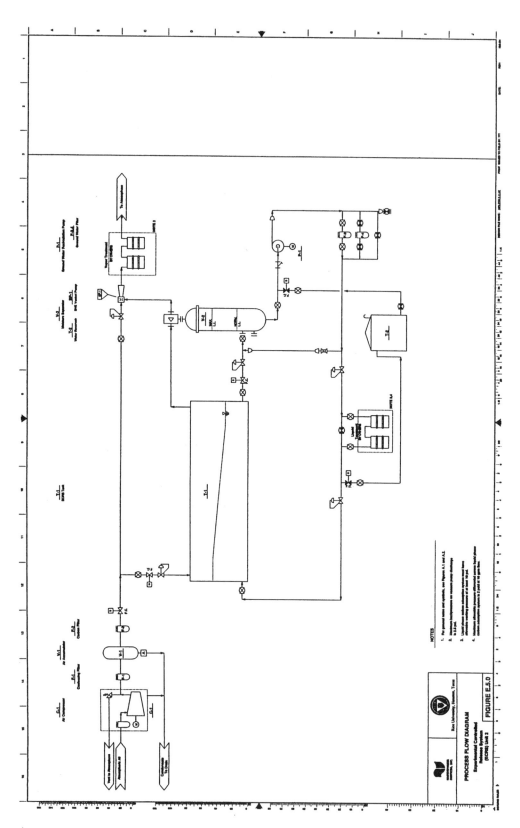

Figure E5.0 Process flow diagram.

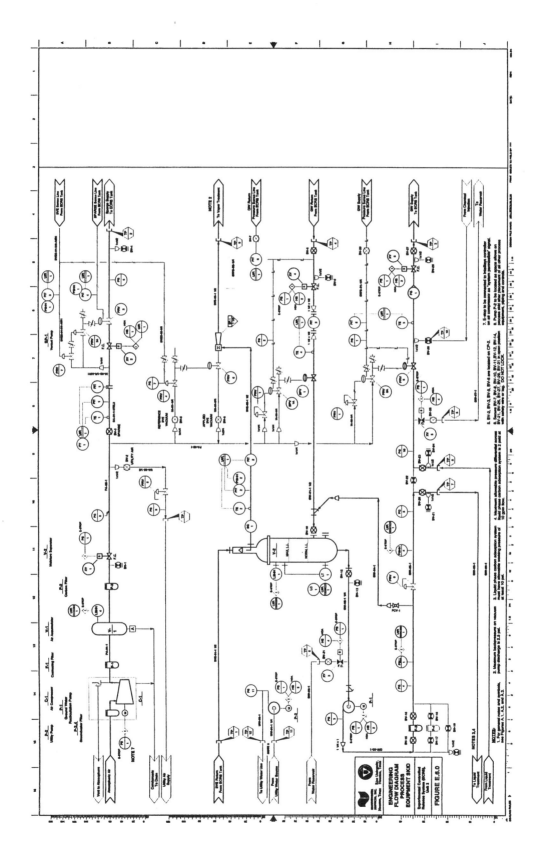

Figure E.6.0   Engineering flow diagram, process equipment skid.

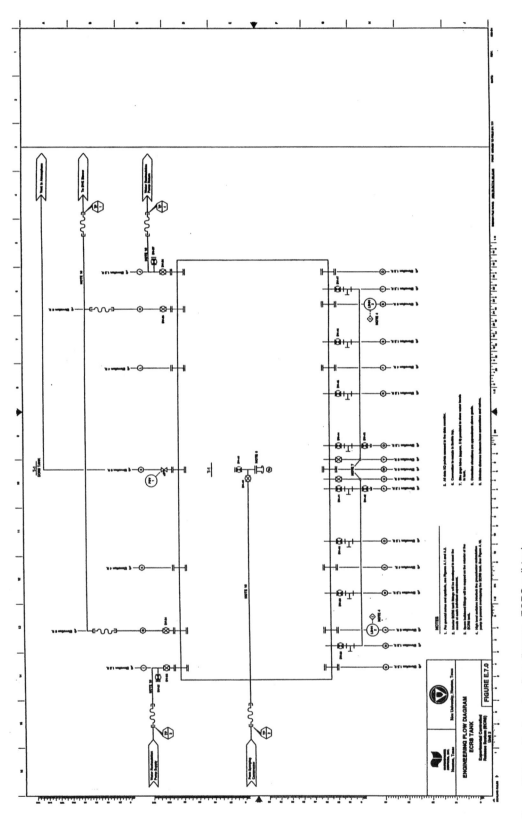

Figure E7.0 Engineering flow diagram, ECRS soil tank.

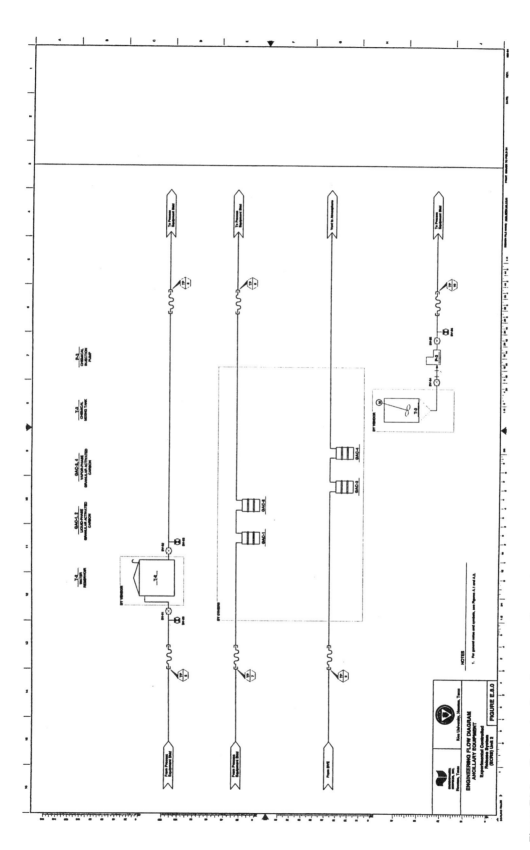

Figure E8.0   Engineering flow diagram, ancillary equipment.

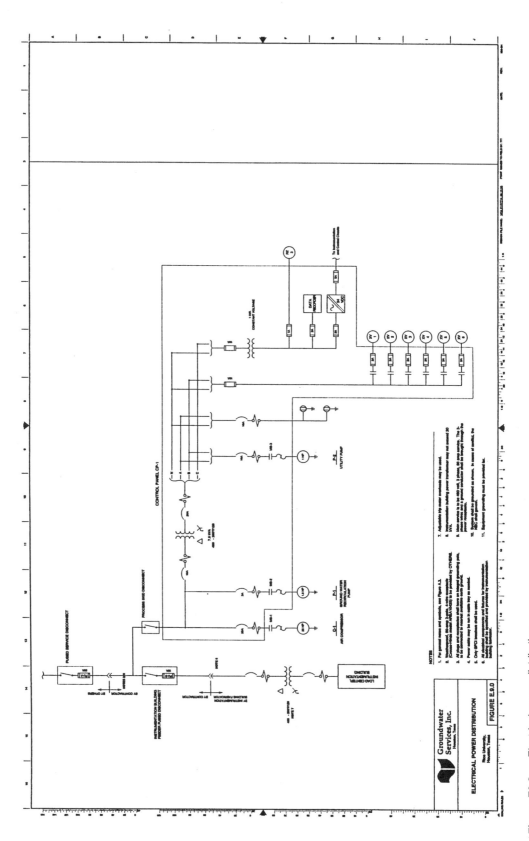

Figure E9.0   Electrical power distribution.

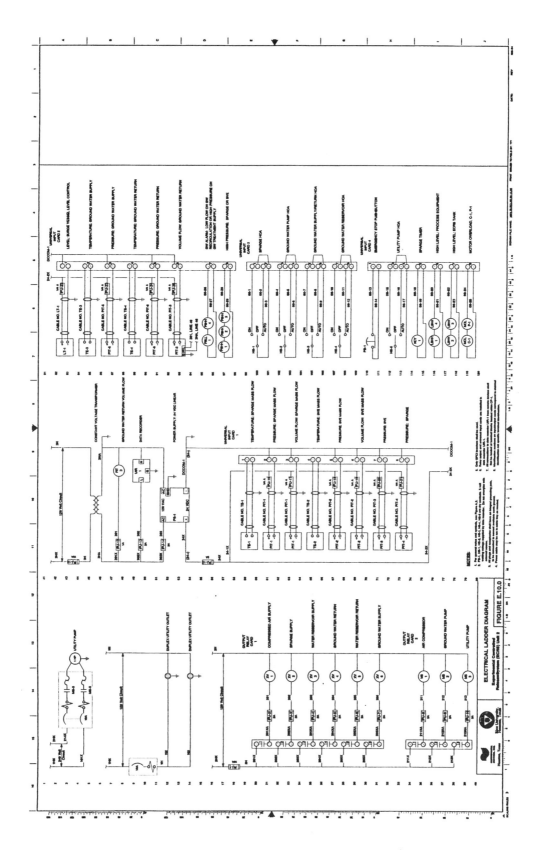

Figure E10.0 Electrical ladder diagram.

References

AATDF Report TR-96-2, 1996. Experimental Controlled Release System: ECRS Unit 1 Development, Energy and Environmental Systems Institute, Rice University, Houston, TX.

AATDF Report TR-97-1, 1997a. Experimental Controlled Release System: ECRS Unit 1 Operations and Maintenance Manual (3 volumes), Energy and Environmental Systems Institute, Rice University, Houston, TX.

AATDF Report TR-97-5, 1997b. Experimental Controlled Release System: ECRS Unit 2 Operations and Maintenance Manual (3 volumes), Energy and Environmental Systems Institute, Rice University, Houston, TX.

Aller, L., Bennet, T., Lehr, R.J., and G. Hackett. 1987. DRASTIC: A Standardized System for Evaluating Ground Water Pollution Potential Using Hydrogeologic Settings, EPA-600/2-87-035, U.S. Environmental Protection Agency, Ada, OK.

Hirasaki, G.J., Miller, C.A., Szafranski, R., Tanzil, D., and J.B. Lawson. 1997. Presentation: Field Demonstration of the Surfactant/Foam Process for Aquifer Remediation, SPE 39292, Soc. Petrol. Eng., Technical Conference and Exhibition, San Antonio, TX.

Johnson, R.L., P.C. Johnson, D.B. McWhorter, R.E. Hinchee, and I. Goodman, 1993. An Overview of In situ Air Sparging, *Groundwater Monitoring & Remediation*, Fall, pp. 127–135.

Knox, R.C., Sabatini, D.A., Harwell, J.H., Brown, R.E., West, C.C., Blaha, F., and C. Griffin. 1997. Surfactant Remediation Field Demonstration Using a Vertical Circulation Well, Ground Water, 35:6, pp. 948–953.

Index